WITHDRAWN

WINNIPEG
MAR 3 1 2017
PUBLIC LIBRARY

D1294593

WITHDRAWN

WINNIPEG
MAR 3 1 2017
PUBLIC LIBRARY

WINNIPEG
MAR 3 1 2017
PUBLIC LIBRARY

ELECTRICAL CODE SIMPLIFIED
Book 1 – Residential

This book is intended for use in all jurisdictions that have adopted the Canadian Electrical Code, including those that have adopted the Code with certain amendments and additions. Where applicable, it's noted in the text that a certain amendment has been made by a particular jurisdiction, and an icon of the applicable province appears in the margin adjacent to the amendment as follows:

An amendment (variation) from the Canadian Electrical Code that affects Ontario. When either this symbol or the one below appears in the margin, watch for details in the adjacent text that indicate variations from the unamended Canadian Electrical Code that affect Ontario or BC. Otherwise, just assume they are following the unamended Canadian Electrical Code.

An amendment applicable to only British Columbia. Watch for details as explained for Ontario, above.

The Canadian Electrical Code is the product of literally thousands of men and women from all across Canada who, since 1927, have generously contributed their expertise, their special training, and their invaluable experience to making this the outstanding document that it is. We can be justifiably proud of our Canadian Electrical Code – we own it, collectively.

The primary purpose of each part of the Code is safety where ever electricity is being generated, transported or used. However, a safety code is of no value until it is well published and understood. Electrical Code Simplified has opened the Code and made it the friend, not the enemy, of a very large number of Canadians who have discovered that they can understand the Code and can apply it to great benefit.

In more than thirty years' experience as an Electrical Inspector, the author often saw the need for a clear and simplified version of our Canadian Electrical Code. This book is intended to fill that need insofar as loomex wiring is concerned. It should be used as a guide only. Where there is any doubt as to interpretation or application, the local Inspector should be consulted.

Every effort has been made to keep the explanations brief, yet clear, and in everyday language. Many illustrations are provided to clarify certain points otherwise difficult to explain. As a further aid to the student or contractor, a large number of references from the Canadian Electrical Code are given. The material is dived into 21 sections, each dealing with a specific aspect of the electrical installation.

It is the sincere hope of the author that this book will be helpful to you, the reader.

Corrections to any errors or omissions found subsequent to the printing of this book are listed at www.psknight.com.

To be notified by email of new editions, please sign up at www.psknight.com.

ALSO AVAILABLE:

ELECTRICAL CODE SIMPLIFIED — COMMERCIAL & INDUSTRIAL
For the Professional Electrician

No jargon Electrical Code Simplified makes electrical law clearly and quickly relatable. It gives straight answers on tough issues.

Code Rules seldom apply in isolation; there are usually other Rules that should be considered before a decision is made. This book collects these different applicable Rules, explains how they relate to each other and how they apply, and where helpful, offers commentary on the background to difficult sections of electrical law.

Available online at
www.psknight.com.

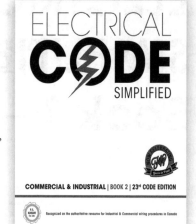

P.S. Knight Co. Ltd.
6423 Burbank Rd. S.E.
Calgary, Alberta
T2H 2E1
Tel: 1.403.648.2908
www.psknight.com

CONTENTS

1.   RESPONSIBILITIES

Some of the requirements of persons doing electrical wiring are as follows:

ELECTRICAL CONTRACTORS

Rule 2-004 - The Electrical Contractor must:

- Obtain permits for all of their work.
- Do all work according to code.
- Not work under a permit issued to the owner or another Contractor.
- Not take out a permit for work done by someone else.
- Not hire uncertified people on a job basis to work under their certificate. These workers are then really contracting without proper Certification.
- Not undertake to do work which exceeds their certification.
- Make sure that all electrical equipment is properly certified before it is connected. See below for a list of acceptable certification marks.
- Declare in writing to the Electrical Inspector when the work is ready for inspection.
- Not cover any wiring until acceptable arrangements have been made with the Inspection Office.

STUDENTS

Students who familiarize themselves with the information contained in this book will find it easier to write the official Government examination for a Certificate of Qualification.

The Government Examiner will supply you with an official Code book, a set of Bulletins, a copy of the Regulations, and all the paper you will need to write the exam. Do not bring your own. Just bring a pen, pencil, a calculator and a clear thinking head.

The student should understand:

- The student must be a mature adult with at least 3 years experience in the installation and maintenance of electrical equipment or have special training satisfactory to the Chief Inspector before they may write the official examination. Documentary proof, required by the Chief Inspector, may consist of a certificate of achievement from a training school, i.e. Apprenticeship certificate or letters from employers stating level of responsibility and training received.
- Failure to obtain a passing mark on the official exam may require a waiting period before a second attempt is permitted. These details vary between provinces.
- Once the exam is passed, the student is issued a Certificate of Qualification. Depending on the Regulations in the province, the certificate permits the holder to engage in a limited amount of electrical work. Processing forms to obtain electrical permits are usually available at any provincial electrical inspection office.
- One more thing - After hours of study and agony over the exam, you face the Inspector in the field with each installation you do. If your work is good and your cooperation is good, you will have no problem. If it is not good, the Board of Examiners may want to talk to you about it - may even want to give you a vacation (suspend your Certificate) for a period of time.
- After two failures, it appears the thing to do is to become interested in some other trade - plumbing maybe.

HOMEOWNERS

The homeowner:

- Must obtain an electrical permit for all electrical work they do, including the wiring for ranges, dryers, furnace, extra outlets, etc. The rules require an electrical permit for any additional wiring, any alteration, and any new wiring. It is like a building permit; it permits the holder to install certain electrical wiring and equipment in their own home. This permit must be obtained before any electrical work is begun.
- May be required to answer a few questions asked by the local Electrical Inspector to show a certain level of competency before a permit is issued but that should not be difficult provided you have read through this book. The Inspector cannot do the planning and layout work for you. The purpose of this book is to walk you through the whole installation and prepare you with good answers so that you should not have any difficulty with the Inspector in obtaining a permit or with their inspection of your work.
- Should note two things about that permit: First, that it has an expiry date. This means that you are expected to complete your work within the time allowed. An extension is usually possible by permission of the Inspector. Second, that you assume legal responsibility for all the electrical work you do. Again, these things are not difficult if you follow the instructions given in this book.
- Permit Holder Must do all the work themselves — they may not allow anyone to do any of the work covered by the permit issued to the owner.
- May do the wiring in their own home only. This is the basis on which a permit is issued to an owner. They may not install the wiring, any wiring, in their own building if it is to be rented or sold. They may obtain an electrical permit only for their own personal dwelling. They must do all the work themselves. No one, other than their immediate family, such as a father, brother, son, may do any of the electrical work under this permit. It has nothing to do with payment for work done. Even an Electrical Contractor could not work under a permit issued to an owner. Treat it as you would your drivers license.

- Must do all the work according to the electrical code.
- Must notify, in writing, the Electrical Inspector when the work is ready for inspection. One must not request an inspection before they are, in fact, ready. The Inspector may charge a re-inspection permit fee if they find the work is not ready or if it has been improperly installed.
- Complete the entire installation including all fixtures, switches, plates, etc. before a final certificate of approval is issued.

2. CSA CERTIFICATION

Rule 2-024 - CSA is no longer the only testing and labeling agency acceptable in Canada. There are now ten certification agencies. Each of these agencies performs prescribed tests to ensure the electrical equipment and appliances sold in this province are manufactured according to a rigid set of standards. Following is a list of the certification marks used by these agencies.

CGA - Canadian Gas Association - This is what to look for on gas appliances such as furnaces, dryers, cooking stoves etc. If this label is in place it usually means both the gas and the electrics are certified for use in Canada.

CSA - Canadian Standards Association - This is the old reliable and very familiar certification label. Until recently, this label, and the provincial label, were the only two labels acceptable in this province. This label is still acceptable just as it always was on almost everything electrical.

ETL - Inscape Testing Service - This label is not as well known here in Canada but we can expect to see it on an ever increasing number of electrical devices. Note the small "c" outside and to the left of the circle. This means that equipment with this label is acceptable for use in Canada. If an electrical device has this label look for the small "c." If it is not there the device is not acceptable for use in Canada.

Warnock Hersey - This certification mark already appears on heating equipment. We can expect to find it on a broad range of electrical devices in the future.

UL - Underwriters Laboratories - This is a major testing agency in the United States. Note the small "c" outside and to the left of the circle. This means that equipment with this label is acceptable for use in Canada. If an electrical device has this label look for the small "c." If it is not there the device is acceptable for use in the United States but it is not acceptable for use in Canada.

ULC - Underwriters Laboratories of Canada - This certification marking is used mainly on fire protection equipment. You will find it on smoke alarms used in single family homes.

The next three approval markings, Met, Entela and O-TL, are from the latest agencies accredited by the Standards Council of Canada to approve electrical equipment for connection and use in Canada. Note that in each case there is a small "c" outside and to the left of the certification mark. Without that small "c" these marks are not acceptable in Canada.

Ontario Electrical Safety Authority. This marking is applied to electrical equipment which for one reason or another must be tested for acceptance in the field. This would include specialty items that have been brought in as settlers effects or items imported on a one time basis. This label indicates the equipment has been examined and found to be acceptable for use in Ontario.

Provincial Government marking. - This marking is applied to electrical equipment which for one reason or another must be tested for acceptance in the field. This would include specialty items which have been brought in as settlers effects or items imported on a one time basis. This label indicates the equipment has been examined and found to be acceptable for use in a given province.

Certification markings are very important for your safety. Each electrical device in your home, from the main service panel to the cover plate on the light switch, must bear one of these markings to show it has been properly tested and is certified. It may be tempting to pick up that cheap, uncertified, light fixture in Mexico or elsewhere, or that under the counter electrical gizmo at a fraction of the cost for the same thing back home but don't do it. It may be a fire hazard or an electrical shock hazard. Manufacturing to a safe standard and submitting that equipment to a laboratory for rigid tests is an expensive process but it is your assurance that device meets certain minimum standards. Always look for one of the above certification marks on all electrical equipment you purchase. If you do not find any of these markings bring the device to the attention of the store manager.

Unfortunately the number of Certification Marks acceptable in Canada today undermines the idea that a home owner will recognise an illegal marking. There are at least ten more markings that are also acceptable in Canada, but those are not shown here. In this author's view, the millions of home owners and the many thousands of workers involved in the application of this regulation in the retail outlets all across Canada cannot possibly be expected to know each of these different safety markings and consequently, this important fundamental safe guard, I believe, is compromised.

3. ELECTRICAL INSPECTIONS

WHEN IS AN INSPECTION NOT REQUIRED?

There are some instances in which a permit is not required. Generally where the work replaces luminaires, receptacles, or general use switches in owner-occuped single dwelling units providing that the work doesn't necessitate a change to the branch circuit box. The best strategy is to check with your local inspection office. Neglecting to do so may nullify your home insurance in the event that a change you make results in an insurance claim.

ROUGH WIRING INSPECTION

Before calling for an inspection of the rough wiring make sure that:

- The electrical service equipment is in place. The service conduit or cable, the meter base, service panel and the service grounding cables should all be installed, where it is practical to do so, for the rough wiring inspection.
  If you are using service conduit, not cable, the conductors need not be installed for the rough wiring inspection, but the conduit, and all its fittings, wherever practicable, must be in place.
- All branch circuit cables are in place and properly strapped and protected from driven nails. The Inspector needs to see the arrangement of the installed branch circuit cables, the size of cables used, and the number of outlets on each circuit. and the joints and splices you have made. Don't forget, all joints, splices and ground wire connections to boxes must be completed as far as possible.
- All cables are properly terminated in boxes and all joints and splices in the boxes are completed as far as possible. Do not connect or install any fixtures or devices. All bond wire connections to outlet boxes must be completed and where pigtails are required they should be in place. The connections in boxes must be complete to the point where all that is left to do is make the terminations to the devices. The Inspector needs to be able to determine correct box size, correct bonding connections and correct circuit connections.

> Your Provincial Government Approval Label

AVOID COSTLY REJECTIONS AND DELAYS

Before calling for inspection of the rough wiring check your work very carefully as an Inspector would do their inspection. Check every detail. If you have followed the instructions in this book you should do well.

Do not cover any wiring, not even with insulation, until it has been inspected and is approved for covering.

In some rural areas a sketch map may be necessary to help the Inspector find your premises. Also, if a key is required, provide instructions on where to find it.

FINAL INSPECTION

For this inspection your installation must be entirely complete. It's a good idea to check it carefully as an Inspector would do their inspection. Use this book as you go through your installation to check off each item. Look for forgotten or unfinished parts. Check for such things as circuit breaker ratings, tie-bars, circuit directory, grounding connections etc. Look for open KO holes or junction boxes and unfinished outlets. Don't forget, if any of the appliances such as a dish washer is not going in just yet you must terminate the supply cable in a fixed junction box, (the box must be fastened in place) and must be complete, with cover, until that appliance is actually installed.

Electrical Permit - Check also if your permit covers all you have installed. The Inspector is one of the good guys. Their concern is that your installation is safe and that it meets minimum code requirements. That is, alter all, exactly what you want too.

4. ELECTRICAL INSPECTION OFFICES

The name of your local Electrical Inspector, their office location, phone number and office hours are not listed here because this book may be used in a number of provinces and territories. Check your telephone directory for the nearest Electrical Inspection office.

The local Electrical Inspector will usually answer any reasonable, and specific, question regarding your installation, however, they should not be expected to design the installation for you. If you encounter a problem for which you need an official interpretation it's usually okay to call your local Electrical Inspector, but not until you have done your research and checked it out carefully in this book. The question should then be put as clearly as possible so that the Inspector can give you a meaningful answer.

The electrical Code simply sets out certain minimum requirements we are expected to comply with in order to have a safe installation. It should not be used as a design manual.

Electrical Code Simplified (this book) is designed to answer most of the questions you will encounter as you plan your electrical installation.

5.   SERVICE SIZE (AMPACITY)

Rule 8-200 - We must begin with the electrical service box. It may appear to be very difficult to install a new electrical service in your house, but it need not be. If you read these instructions carefully it should be easy and enjoyable and what's more, you should save a bundle.

The rules permit a service as large as we like but not as small as we like. There is a definite minimum size we must have if it is going to be passed by the Inspector.

Service size is based on two things: calculated load, and minimum service size.

CALCULATED LOAD

This is the sum of all the loads after certain demand factors are applied. For an average 90 m^2 (968 sq. ft.) house with electric range and dryer but with gas or oil heating, the service demand load consists of 5,000 watts for what the Code calls basic load, plus 6,000 watts for the range plus another 1,000 watts for a 4,000 watt dryer. Total is 12,000 watts. To find amperes we need to divide the 12,000 watts by the service voltage. 12,000 watts divided by 240 volts is 50 amps. Don't forget this is calculated amperage not minimum service size and there is a big difference.

NOTE ABOUT ALUMINUM CONDUCTORS

The Canadian Electrical Code allows service conductors and neutral conductors to be either copper or aluminum. However, aluminum conductor installation requires great care. This special procedure is not covered in this book, so aluminum is not mentioned in the material lists or discussion that follow.

MINIMUM SERVICE SIZE (AMPACITY)

This is based on floor area.

- 60 amp. For any house with less than 80 m^2 (861 sq. ft.) floor area. This includes the areas of all the floors except the basement. Basement floor area is ignored completely for this purpose.
- 100 amp. For any house with 80 m^2 (861 sq. ft.) or more floor area. As above, this includes all floors except the basement.
- Check Table 39 (below) to see if you can use a smaller service size than the size based on floor area.

Even if the actual load is, say 30 amps, the minimum size service permitted is 100 amps if the floor area is 80 m^2 or more. This extra capacity is for future load.

The "Service Ampacity Table for Homeowner Use" on p. 10 has been developed to simplify the service size calculation. It shows a progressive load - in other words, any service size given in any column assumes all the loads shown above that value are going to be used.

For example, a 90 m^2 (no basement) house with no electrical appliances such as a range, dryer, water heater or electric heating would require a 100 amp service. The material required for this service is listed under B, p. 5.

If the 90 m^2 no basement house had a 12,000 watt range, 4,000 watt dryer, 3,800 watt water heater, the service size would still be only 100 amps. This is the minimum size service for this floor area. The material required for this service is listed under B, p. 5.

If the 90 m^2 no basement house had a 12,000 watt range, a 4,000 watt dryer, a 3,800 watt water heater and a 15,000 watt electric hot water boiler, the service size would need to be minimum 117.1 amps. The material required for this service is listed under C, p. 8.

Once you have determined the correct list of material for your house, measure the lengths required for your job. You can use the following lists when shopping for the materials you will require. Note: The material lists are based on floor area and the "Service Ampacity for Homeowner Use" table. However, if you calculate your ampacity based on loading, Table 39 may permit you to use a smaller conductor than that shown in the material list.

Table 39
MINIMUM PERMITTED SIZE FOR 3-WIRE 120/240 V AND 120/208 V SERVICE CONDUCTORS FOR SINGLE DWELLINGS*

Service or Feeder Rating, A	Copper (See Note 3)		Aluminum (See Note 3)	
	For Calculated Loads up to, A*	Conductor Size	For Calculated Loads up to, A*	Conductor Size
60	60	6 AWG	53	6 AWG
100	89	4 AWG	95	2 AWG
125	121	2 AWG	125	1/0 AWG
200	184	2/0 AWG	189	4/0 AWG
225	210	3/0 AWG	215	250 kcmil
400	352	400 kcmil	357	600 kcmil

These ampacities are the ampacities given in Tables 2 and 4 for 75 °C, increased by 5% in accordance with Rule 8-106(1). See note 2.

NOTES

(1). This table applies only to conductors sized in accordance with rules 8-200(1) and (2) and 8-202(1).

(2). The 5% allowance in rule 8-106(1) cannot be applied to these values as the 5% allowance has already been added to the applicable 75 °C ampacities from tables 2 and 4.

(3). If the calculated load exceeds the limit shown in the table, the next larger size conductor shall be used.

(4). *The full title of Table 39 in the Code is "Minimum Permitted Size for 3-Wire 120/240 V and 120/208 V Service Conductors for Single Dwellings and Feeder Conductors Supplying Single Dwelling Units of Row Housing of Apartment and Similar Buildings and Terminating on Equipment Having a Conductor Termination Temperature of Not Less Than 75 °C"

LIST OF MATERIALS

The following lists assume that you will use copper conductors, not aluminum. Remember you need twice as much hot conductor length as neutral conductor.

A. SERVICE SIZE - 60 AMPS

This service size may be used only if the total floor area, above ground, is less than 80 m^2 (861 sq. ft.).

Service switch, fuse or breaker rating .... 60 amps
Hot conductors.. 2 - #6 R90XLPE copper (black, red or blue)
Neutral conductor (See Note 4 below)... 1 - #6 R90XLPE copper (white or natural grey)
Service raceway conduit .... 1 in. or use #6 copper TECK cable
Meter base rating (See Note 4 below)... 100 amps
Service grounding conductor .... #6 (or larger) bare copper, see p. 45 for details
Service panel size (See Note a below and Note 4) .... 16 circuits (minimum)

Note a: This panel may supply all the normal outlets and a central gas or oil furnace in this small house.

B. SERVICE SIZE - 100 AMPS

• See table, on p. 10 for floor area

Service switch, fuse or breaker rating .... 100 amps
Hot conductors .... 2 - #3 R90XLPE copper (black, red or blue)
Neutral conductor (See Notes 4 and 6)... 1 - #6 R90XLPE copper (white or natural grey)
Service raceway conduit (See Note b below) .... 1-1/4 in. or use #3 copper TECK cable
Meter base rating (See Note 4 below)... 100 amps
Service grounding conductor .... #6 bare copper, see p. 45 for details
Service panel size (Notes a and 4 below) .... 24 circuits, see p. 37 for details

Note a: This panel may supply all the normal outlets in the house and a central electric furnace, an electric boiler or baseboard heaters.

Note b: A 1 in. (27) would be acceptable for this 100 ampere service but the list above shows 1-1/4 in. conduit. A 1-1/4 in. conduit will permit a service upgrade to 150 amps at minimum expense later, when large loads are added.

C. SERVICE SIZE - 120 AMPS

See Table, p. 10 for floor area

Service circuit breaker rating ....125 amps w
Hot conductors ....2 - #2 R90XLPE copper (black, red or blue)
Neutral conductor (See Note 4 below) ....1 - #6 R90XLPE copper (white or natural grey)
Service raceway conduit (See Note b above) ....1·1/4 in. (35) or use #2 copper TECK cable
Meter base rating (See Note 4 below) ....200 amp
Service grounding conductor ....#6 bare copper, see p. 45 for details
Service panel size (See Notes a and 4 below) .... 24 circuits, see p. 37 for details

Note a: This panel may supply all the normal outlets in the house and a central electric furnace or electric boiler. If heating is with electric baseboards you will need more than 24 branch circuits, see list D below.

D. SAME AS ABOVE EXCEPT

Service Panel Size - see p. 37 for details. 30 circuits (See Note a below)

Note a: This panel may supply all the normal outlets in the house and a central electric furnace, an electric boiler or electric baseboard heaters.

E. SERVICE SIZE - 150 AMPS

See Table, p. 10 for floor area

Service circuit breaker rating .... 150 amps
Hot conductors ....2 - #1/0 R90XLPE copper (black, red or blue)
Neutral conductor (See Note 4 below) ....1 - #6 R90XLPE copper (white or natural grey)
Service raceway conduit ....1-1/4 in. (35) or use #1/0 copper TECK cable
Meter base rating (See Note 4 below) .... 200 amp
Service grounding conductor ....#6 bare copper, see p. 45 for details
Service panel size (See Notes a and 4 below) .... 30 circuits, see p. 37 for details

Note a: This panel may supply all the normal outlets in the house and a central electric furnace or electric boiler. If heating is with electric baseboards the 30 circuit panel board is too small. You will need to install a 40 circuit panel board. See Minimum Circuits Required, p. 40.

F. SAME AS ABOVE EXCEPT

Service Panel Size, see p. 37 for details .... 40 circuits. (See Note 4 below)

This panel may supply all the normal outlets in the house and a central electrical furnace, an electric boiler or electric baseboard heaters.

G. SERVICE SIZE - 200 AMPS

Service circuit breaker rating .... 200 amps. (See Note 3 below)
Hot conductors ....2 - #3/0 R90XLPE copper (black, red or blue)
Neutral conductor (See Note 4 below) ....1 - #4 R90XLPE copper (white or natural grey)
Service raceway conduit ....1·1/2 (41) or use #3/0 copper TECK cable
Meter base rating (See Note 4 below) ....200 amps
Service grounding conductor .... #6 bare copper. See p. 45 for details
Service panel size (See Note a and 4 below) .... 30 circuits, see p. 37 for details

Note a: This panel may supply all the normal outlets in the house and a central electric furnace or electric boiler. If heating is with electric baseboards the 30 circuit panel board is too small. You will need to install a 40 circuit panel board. See "Minimum Circuits Required," p. 40.

NOTES FOR THE INSTALLER

(1). Minimum Panel Size - The above lists give the minimum panel size required by code in each case, however, this may not be enough for your installation. Make sure that you have a sufficient number of circuit spaces available in your panel. See also under (c) "How Many Circuits Do I Need," p. 40 and the chart, p. 40.

(2). Spare Circuits - 2 Required - Rule 8-108(2) requires that there be at least two spare circuits left in the panel after you have connected all the circuits you have installed. These two circuits are for future use. See also (c) "How Many Circuits Do I Need," p. 40.

(3). 200 Ampere Service - The footnotes for Table 2 in the Code, permit #3/0 90° copper conductors to be used for a 200 amp service as shown in the "List of Materials" above. This smaller conductor permits 1-1/2 conduit to be used for a 200 ampere service.

(4). Neutral conductor terminations in the Meter Base and the Service Panel - Ontario Bulletin 4-3-5 requires that the neutral service conductor be the same size as the hot conductors for 100 amp service. Where the Code permits a small neutral service conductor as shown in the "List of Materials" starting on p. 5, this may be a problem when you are shopping for both the service meter base and service panel. The terminal lugs for the neutral conductor in the meter base and the service panel must be properly rated for the smaller s permitted under the Code. Look for the conductor size marked on the connector lugs. Do not attempt to make any connections with connector lugs that are improperly rated for the conductors you are installing. Installing a larger neutral conductor may add to the cost but it could also resolve the problem. Don't forget, if you install a larger service neutral conductor you may also need to use the next size larger service conduit. Manufacturers of meter bases and service panels, no doubt, know this detail and they are making changes that the Code now permits a smaller neutral, and they are providing properly rated connector lugs to agree with the Rules, but old stock may still be available. Careful shopping could save a lot of trouble.

(5). Bare Neutral - Rules 4-020 & 6-308 - Each list of material refers to an insulated neutral conductor, however, the rules do not require the neutral to be insulated; it may be bare. Using a bare neutral may save a bit of money and it is easier to form into the desired shape to make connections. Note that a bare neutral, if of copper, may be run in PVC conduit or in EMT.

Where a bare neutral enters a meter base, or switch, or a panel, it must be insulated to prevent contact with live parts. In most cases it should be possible to train this bare conductor so that there is no danger of inadvertent contact with live parts. However, where it is not possible to prevent inadvertent contact with live parts, the bare conductor must be insulated. Apply a layer of electrical tape throughout the length of the exposed section. This layer of tape should be equal to the thickness of the insulation on the hot conductors.

NOTES FOR THE STUDENT

(1). 60 amp service - In this case Table 2 permits the #6 R90XLPE copper hot service conductors to carry 65 A. That is 5 A more than we need, but if we use these service conductors we must enter Table 16A/B with 65 amp service conductors, not 60 amp. In that case Table 16A/B requires a #8 copper R90XLPE neutral as shown in the "List of Materials" on p. 5.

(2). 100 amp service - In this case the #3 R90XLPE service conductors have a maximum ampacity of **100 A**. Table 16A/B requires a #6 neutral as shown in the List of Material.

(3). Service neutral conductor size - Ontario Bulletin 4-3-5 The requirements for the service neutral size was changed The size of conductor required for the service neutral in a single family house was reduced in the 20th edition of the Canadian Electrical Code. This change was a long time coming and it rocked the sensitivities of many all across Canada, but, actually, it did so in a comfortable and considerate way. A smaller neutral may also mean a smaller service conduit and that translates into real savings. There is no point in installing a large service neutral which is rated far above what is actually needed or used at any time in the life of that installation. There may be some misunderstanding of the purpose for the neutral service conductor. Some may be concerned about the much smaller size of this conductor permitted now by the new Ontario Safety Code Electrical Code. The following will help to clarify the purpose of the neutral in terms of actual ongoing load and it's very important purpose in short circuit conditions.

The 20th edition of the Canadian Electrical Code was the first Code to permit a much smaller service neutral conductor than was possible under any of the old Codes. Please note, the British Columbia Safety Authority did not revise this; they adopted it as it is applied above in the List of Materials, p. 5. Rules 4-022 and 10-204 were accepted without any provincial amendments. You see, the Canadian Electrical Code Committee Engineers and the IEEE Engineering Standards recognized the fact that the previous Code Rules required a larger neutral and service conduit than is actually necessary. They also recognized the fact that the larger neutral represents a huge waste of copper, insulating material on the conductors and steel in the conduit. The additional, or extra ampacity, required by the old Code is unnecessary for safety for the present load, or for future load additions and therefore much smaller conductors are permitted by the legally adopted Rules.

There are two very practical reasons why the neutral conductor is so important. First, according to Rule 4-024(1) this conductor must be large enough to carry the unbalanced current between the two hot service conductors. Second, according to Rule 10-204(2)(b) it may not be smaller than required by Table 16A or B. Table 16A or B is used for minimum sizing of bonding and grounding conductors to provide adequate ampacity for all grounding and bonding conductors in the Code.

The application of this Table ensures that our neutral conductor size is large enough to carry sufficient current to open the service breaker in the event of a short circuit anywhere in the system. These are the only two reasons for that neutral conductor. The calculation below demonstrates that these smaller neutral conductors really are adequate according to Code. It is very important that these two rules are satisfied.

A Directive cannot modify a Rule. This fundamental principle is clearly made in the following statement A directive is a legal document with authority under the Safety Standards Act but it cannot amend a code rule and where a conflict exists, the code regulation (the Rule) takes precedence.

The first reason for the neutral conductor - Rule 4-024(1) - It must be adequate to carry the unbalanced load: The Calculation of the neutral load in a 180 m^2 house looks like this:

	CONNECTED LOAD	CALCULATED LOAD	NEUTRAL LOAD
First 90 m^2 (These are all 120 volt loads)	5,000 watts	5,000 watts	5,000 watts
Next 90 m^2 (Total floor area = 1938 sq ft.)	1,000 watts	1,000 watts	1,000 watts
First range	12,000 watts	6,000 watts	0 watts
Plus 25% of 12 kw for a second range	12,000 watts	3,000 watts	0 watts
Dryer, 25% of 4,000 watts	4,000 watts	1,000 watts	0 watts
Water heater, 25% of 3,000 watts	3,000 watts	750 watts	0 watts
Electric heating & sauna is 11 kw. plus 3 kw. = 14 kw.			
First 10 kw. at 100%	10,000 watts	10,000 watts	0 watts
Balance, 4 kw. at 75%	4,000 watts	3,000 watts	0 watts
Totals watts	51,000 watts	29,750 watts	6,000 watts
Total amps	212.5 amps	123.9 amps	25.0 amps

The example above shows that the neutral load is only 6,000 watts, or 25 amps, for this medium sized house while the actual (total) connected load in the house is 51,000 watts. Remember that the neutral carries only the unbalanced 120 volt loads. It is completely unaffected by any of the 240 volt loads as the calculation above shows. Note the zeros in the third column. The Rule allows for a 120 volt unbalance between the hot conductors.

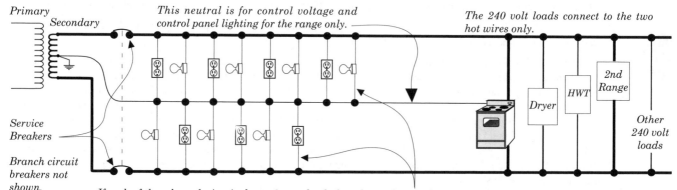

If each of these branch circuits has a 2 amp load, then the top hot conductor would carry 16 amp and the lower hot conductor would carry 12 amps and the neutral would carry the difference, or 4 amps. These loads are constantly changing but the neutral carries only the unbalanced load.

What is often forgotten is the fact that the neutral carries only the unbalanced load and that any balanced 120 volt loads actually operate as 240 volt loads. The illustration above shows a hypothetical load from the Hydro transformer through the service breakers to eight branch circuits on the top hot conductor and six branch circuits on the lower hot conductor (branch circuit breakers are not shown). Each circuit will normally have a different current load but if, for example, we say each 120 volt circuit has a 10 amp load, then the balanced portion of the 120 volt loads is 60 amps on each hot conductor and nothing at all in the neutral conductor. The balanced portion is operating as a 240 volt load without affecting the neutral conductor. However, there are two more 10 amp loads on the upper hot conductor that are not balanced with a similar load on the lower hot conductor and this is the "unbalanced" current that the neutral must be able to carry safely. The unbalanced neutral load, in this example, is 20 amps. Normally only a few lights and only a few plug outlets are in use at any time. Some plugs are seldom used and some are used frequently. When the loads connected between the top hot conductor and the neutral are equal to the loads connected to the other hot conductor and the neutral, the loads are balanced and there is no current at all in the neutral. When they are not balanced the neutral carries only the unbalanced part, the difference between the top and bottom hot conductors.

Future load additions - Adding another 90 m^2 to the house would increase the neutral load by only 4.1 amps. Adding a 6,000 watt sauna, or 10,000 watts of electric baseboard heaters would add connected load, and calculated load, but would not affect the neutral load at all as the calculation on the previous page shows.

The Second reason for the neutral conductor is Rule 10-204(2)(b)- The neutral conductor must be capable of carrying sufficient fault current to quickly open the main service fuse or breaker and this is governed by Table 16A/B for almost all loads referred to in the Code. The List of Materials, p. 5 shows that a #6 copper conductor may now be used for the neutral in a 60 amp house service where the ampacity of the hot conductors is no more than 60 amps. A #6 copper conductor may be used for the neutral in a 200 A service in a single family house as shown in the List of Material, p. 5.

A #6 bare copper grounding conductor for a 60 amp service does not affect the size of the neutral conductor provided the neutral (the grounded conductor) is properly sized. It can be smaller than the service grounding conductor when the service grounding conductor is large because its size permits it to be run exposed, Rule 10-806(2). This may be a little confusing because Rule 4-024(3)(b), which required them to be the same size in the old Code, was changed in the new Code but was not marked as having been changed in the new Code.

Yes, Tables 16A and 16B are for bonding conductor sizes. The new Code now regards the service neutral conductor as both a neutral and a bonding conductor because it is actually required to function as both of them. Under normal conditions it functions as the service neutral but under fault conditions it must function as a bonding conductor. The Code is concerned that this conductor is large enough so that it can deliver sufficient fault current to trip the service circuit breaker, or to blow the service fuse in the shortest time possible in the event of a short circuit.

A final thought on this neutral thing - While there is general agreement in the electrical trade that a full size neutral conductor is wasteful because it cannot possibly ever be used, there is disagreement on how much reduction should be permitted. The explanation given will help the student understand the technical reasons why the Canadian Electrical Code will permit the service neutral to be considerably smaller than the hot service conductors.

Important Note about the Following Table and New Table 39

The table on the following page offers one way to find out what ampacity of service conductor your house needs.

In the following table, the ampacity is based on floor area. Now, in the 23rd edition, the Code offers a second way to figure out the required ampacity - that is, using Table 39 on p. 5 of this book.

To use Table 39, you calculate the ampacity based on loading (not floor area), and then you look up the required conductor size in Table 39. The required conductor size from Table 39 is often smaller than the conductor size based on the following table, so it pays to check.

Example:

For a calculated ampacity of 83 A, Table 39 states a that you can use #4 copper conductors, whereas #3 conductors would be required based on the following table.

Service Ampacity Table for Homeowner Use

Connected Load see notes on adjacent page	Floor area without basement is 80 m² / Basement floor area is 80 m² (861 sq. ft.) No basement house	With any size basement	Floor area without basement is 90 m² / Basement floor area is 90 m² (969 sq. ft.) No basement house	With any size basement	If total floor area is between 90 m² and 180 m² (1937 sq. ft.) See Notes 2 & 5	If total floor area is between 180 m² and 270 m² (2906 sq. ft.) See Note 2	If floor area is between 270 m² and 360 m² (3875 sq. ft.) See Note 2
Basic Load only See Note 3	60 Amps A	60 Amps A	100 Amps B	100 Amps B	See Notes 2 & 5	See Note 2	See Note 2
Plus range (up to 12 KW rating)	60 Amps A	60 Amps A	100 Amps B	100 Amps B	100 Amps B	100 Amps B	100 Amps B
Plus 4 KW dryer	60 Amps A	60 Amps A	100 Amps B	100 Amps B	100 Amps B	100 Amps B	100 Amps B
Plus 3 KW water heater See Note 8	60 Amps A	60 Amps A	100 Amps B	100 Amps B	100 Amps B	100 Amps B	100 Amps B
Plus heating with electric hot air furnace or electric hot water boiler							
10 kw	94.7 amps B	98.9 amps B	94.7 amps B	98.9 amps B	98.9 amps B	*103.1 amps C	107.2 amps C
15 kw	110.4 amps C	114.5 amps E	110.4 amps C	114.5 amps E	114.5 amps E	*118.7 amps E	122.9 amps E
18 kw	119.7 amps E	123.9 amps E	119.7 amps E	123.9 amps E	123.9 amps E	128.1 amps E	132.2 amps E
20 kw	126 amps E	130.2 amps E	126 amps E	130.2 amps E	130.2 amps E	134.3 amps E	138.5 amps E
24 kw			*138.5 amps E	*142.7 amps E	*142.7 amps E	146.8 amps E	151 amps G
27 kw			147.9 amps E	152 amps G	152 amps G	156.2 amps G	160.4 amps G
30 kw					161.4 amps G	165.6 amps G	169.7 amps G
See Note 4 for the meaning of the asterisks							
Or if using Baseboard heaters, sum of all heater ratings							
4 kw	69.7 amps B	73.9 amps B	69.7 amps B	73.9 amps B	73.9 amps B	78.1 amps B	82.2 amps B
5 kw	73.9 amps B	78.1 amps B	73.9 amps B	78.1 amps B	78.1 amps B	82.2 amps B	86.4 amps B
6 kw	78.1 amps B	82.2 amps B	78.1 amps B	82.2 amps B	82.2 amps B	86.4 amps B	90.6 amps B
7 kw	82.2 amps B	86.4 amps B	82.2 amps B	86.4 amps B	86.4 amps B	90.6 amps B	94.7 amps B
8 kw	86.4 amps B	90.6 amps B	86.4 amps B	90.6 amps B	90.6 amps B	94.7 amps B	98.9 amps B
9 kw	90.6 amps B	94.7 amps B	90.6 amps B	94.7 amps B	94.7 amps B	98.9 amps B	*103.1 amps D
10 kw	94.7 amps B	98.9 amps B	94.7 amps B	98.9 amps B	98.9 amps B	*103.1 amps D	107.2 amps D
11 kw	97.9 amps B	*102 amps D	97.9 amps B	*102 amps D	*102 amps D	106.2 amps D	110.4 amps D
12 kw	*101 amps D	*105.2 amps D	*101 amps D	*105.2 amps D	*105.2 amps D	109.3 amps D	113.5 amps D
13 kw	*104.1 amps D	108.3 amps D	*104.1 amps D	108.3 amps D	108.3 amps D	112.5 amps D	116.6 amps D
14 kw	107.2 amps D	111.4 amps D	107.2 amps D	111.4 amps D	111.4 amps D	115.6 amps D	119.7 amps D
15 kw	110.4 amps D	114.5 amps D	110.4 amps D	114.5 amps D	114.5 amps D	118.7 amps D	*122.9 amps F
16 kw	113.5 amps D	117.7 amps D	113.5 amps D	117.7 amps D	117.7 amps D	*121.8 amps E	*126 amps F
17 kw	116.6 amps D	*120.8 amps F	116.6 amps D	*120.8 amps F	*120.8 amps F	*125 amps F	129.1 amps F
18 kw	119.7 amps D	*123.9 amps F	119.7 amps D	*123.9 amps F	*123.9 amps F	128.1 amps F	132.3 amps F
19 kw	*122.9 amps E	127 amps F	*122.9 amps E	127 amps F	127 amps F	131.2 amps F	135.4 amps F
20 kw	126 amps F	130.2 amps F	126 amps F	130.2 amps F	130.2 amps F	134.3 amps F	138.5 amps F

This Table shows a progressive increase in the minimum service size required in each case as the floor area and appliance load increases. The letter after each ampere rating indicates which list of materials to use. For Sauna Heaters, see Note 10 on the adjacent page.

NOTES FOR SERVICE SIZE TABLE

(1). Remember, this table gives the minimum service ampacities permitted under the rules. Services of higher rating may be installed and sometimes may be an advantage for future load additions.

(2). The floor areas in all these columns must include the floor area of an in-house garage, (if there is one) but does not include an open carport. Calculate your floor area by adding basement floor area at 75% and the other floor areas at 100%.

(3). Basic load - This includes:
   • All lighting outlets
   • All 15 amp plug outlets
   • Hot air furnace (standard) gas or oil burning type.
   • Any appliance of less than 1500 watts (this is is the same as 12.5 amps at 120 volts) each. This includes loads such as a garburator, freezer, toaster, etc. but does not include fixed (permanently wired not plug in type) electric heating.

(4). * Some amperages shown on the table are marked with an asterisk. This means that the next smaller standard size service is acceptable in this case because it is within 5% of the required minimum size shown. Please note, special permission is not required. This is your choice now under the Code, Rule 8-106(1). Therefore the ampacity of the service switch, or circuit breaker, and the branch circuit panel, may be reduced to the next smaller standard size but the number of the branch circuit positions in that panel must be as indicated, it may not be reduced. For example if you require a 105.2 amp service you could use a 100 amp service but the total number of branch circuit positions in your one or more panels must be at least 24.

For 3-wire 120/240 V and 120/208 V service conductors for single dwellings, or feeder conductors supplying single dwelling units of row housing of apartment and similar buildings, and that are sized according to Rules 8-200(1), 8-200(2), and 8-202(1), the allowable ampacity for copper and aluminum conductors with 90°C insulation that terminate on equipment with a maximum conductor termination temperature of 75°C is as follows:

Copper	Aluminum
No. 2 AWG - 125 A	No. 6 AWG - 60 A
No. 2/0 AWG - 200 A	No. 2 AWG - 100 A
	No. 1/0 AWG- 125 A
	No. 2/0 AWG - 150 A
	No. 4/0 AWG - 200 A
	No. 700 kcmil - 400 A

(5). If your floor area is 90.1 m^2 we must use the column marked "90 m^2 to 180 m^2."

(6). The table may not show the exact rating of your electrical load - in that case you take the next larger ampacity.

(7). Range Plug - Unless you are installing built-in range units you must install a 14-50R 3-pole, 4-wire grounding receptacle for an electric range, Rule 26-744(4). If you plan to use a free standing gas range, the plug for the electric range is still required in the space behind the gas range.

(8). Motor loads - A hot air furnace requires a fan motor, and a hot water heating system requires a motor to circulate the heated water. These motors are small, they require approximately 7 amps at 120 volts. This is included in the values given in the table.

(9). 6 kw Water Heater - This does not include an electric water heater for a hot tub or spa. Most water heater tanks are equipped with 2 - 3 kw heater elements. While this is 6,000 watts in total, the switching arrangement in the thermostat is such that only one 3 kw element is working at any one time. It is a flip-flop switching arrangement. Under normal water use the lower 3 kw element will heat the water. When the demand for hot water becomes too great for the lower element the thermostat disconnects the lower heater element and connects only the upper heater element. The upper element will heat the water in only the top part of the tank, this provides rapid recovery of hot water. When the demand for hot water decreases the thermostat will switch back to the lower heater element again.

(10). 7,600 Watt Water Heaters - This tank has two 3,800 watt heater elements, total wattage is 7,600. With the flip-flop thermostat, as described above, the maximum load on this tank is 3,800 watts. This is 800 watts more than allowed for in the table. For service calculations we add only 25% of this 800 watts. The additional load is only 200 watts. In an exam this is important - even this little bit - but on the job, well, it's not even one ampere more.

Swimming pool, Hot tub or spa - Electric water heaters for these loads are not included in the table. These must be added at 100% of their rating, Rule 8-200(1)(a)(5).

(11). Sauna (Electric heating)- Rule 62-102 says this must be added as fixed electric space heating.

If you are using gas or oil (not electricity) to heat your house but sauna water is heated by electricity, enter the table with the full rating of this sauna heater as if it were an electric baseboard heater. For example, if you are installing a 4.5 kw sauna you would enter the table as if it was a 5 kw baseboard heater then read across under the correct floor area. Baseboard heaters are listed in the lower left corner of the table.

If you are also using electric baseboard heaters as well as a sauna, then simply add the sauna kw load to the total baseboard heater load before you enter the table.

If your are also using an central electric hot air furnace as well as a sauna use the table for all the other loads, then add the sauna load. All the values given in the table are in amperes, therefore, we need to convert the sauna load to amps as well. Do this by dividing the sauna watts by 240 volts (make sure the sauna heater is connected for 240 volts) then add this to the other load amps. Total amps is the minimum service size. Then use the List of Materials, p. 5.

(12). Built-in Vacuum Cleaning System - Most of these systems will draw 12.5 amps or less at 120 volts. In that case they are included in the basic load shown in the table. Those units which draw more than 12.5 amps at 120 volts must be added at 25% of their rating but remember to take only half of the amperage because the service is calculated at 240 volts and the vacuum motor is connected for 120 volts. For example, if your service should be 125 amp according to the above table and you want to add a 14.0 amp vacuum cleaner system, it would look like this:

Other load		125.00 amps
Vacuum system	25% of 14 divided by 2 = 0.25 X 7 =	1.75 amps
	Total =	126.75 amps

Hardly worth the effort but it could mean that the next size larger service conductors may need to be used.

Now we know how large the service must be. For the next step, refer to the List of Materials, p. 5.

DETAILED CALCULATION FORM (FOR STUDENT USE)

Fill in the blanks as required to describe your installation. This calculation gives minimum size service required. Then refer to the appropriate List of Materials, p. 5.

STEP 1: BASIC LOAD

RULE 8-200(1)

1st. 90 m^2 floor area.   = 5,000 watts

Next 90 m^2 floor area or portion thereof .... (Add 1,000 watts).. =_____ watts

Next 90 m^2 floor area or portion thereof .... (Add 1,000 watts).. =_____ watts

Note 1: Floor area in this case must include 75% of the basement floor area plus 100% of all other living floor areas on all floors. All floor areas are inside, actual, floor area measurements.

Note 2: This basic load includes all lighting and plug outlet loads. It includes oil or gas furnace and any other appliances such as built-in vacuum systems (which are rated 12.5 amps or less), swimming pool pump motors, most workshop motors, compactor motors, garburators, air conditioners, each individually rated at not more than 1500 watts (this is 12.5 amps at 120 volts), but does not include any fixed electric space heating.

STEP 2: APPLIANCES

RULE 26-744(4) ALWAYS REQUIRES AN ELECTRIC RANGE OUTLET UNLESS COOKING UNITS ARE BUILT-IN TYPE

Range (For a 12 kw or smaller range) - add 6,000 watts .... =_____ watts

Plus - (If it is greater than 12 kw,) add 40% of that part which is in excess of 12 kw .... =_____ watts

6,000 watts is not a percentage of the range rating - it would be 6,000 watts for any size range up to 12 kw.

2nd. Range - Add 25% of its wattage. (25% of 12,000 for a 12 kw range) .... =_____ watts

(See Appendix B to Rule 8-200 in your Code book.)

Dryer - Add 25% of its rating if a range is provided for .... =_____ watts

Water heater - Add 25% of rating if a range is provided for... =_____ watts

Tankless water heater - add 100% of its rating .... = _____ watts

If this is all the load we have, i.e. if heating is with gas or oil and there is no other large load such as electric sauna etc., then we must determine minimum service size here as follows:

(a.)   If the floor area is less than 80 m^2, this includes the area of all the floors but does not include the basement. Basement floor area is ignored completely for this determination. Then - minimum service size must be 60 amps. See List of Material, p. 5.

(b.)   If the floor area is 80 m^2 or more, for this determination, as in (a) above, we may ignore completely the basement floor area. Then - minimum service size must be 100 amp. See List of Material, p. 5.

STEP 3: SAUNA

THE CODE CALLS THIS SPACE HEATING

Add sauna load at 100% to other loads if house heating is not with electricity .... =_____ watts

If the house is electrically heated see under specific type of heating below.

STEP 4: ELECTRIC BASEBOARD HEATING & SAUNA

Baseboard heaters - add watts of all heaters =_____ watts
Sauna heater =_____ watts

Total =

    _____ watts

1st. 10 kw must be added at 100% (Rule 62-118) .... =_____ watts
All the balance may be added at 75% .... =_____ watts

Total =_____ watts

Thus far all our calculations are based on watts because that is what many of the demand factors in the Code are based on. The balance of our calculations, however, are based on amperes, not watts. At this point it is best to convert the total watts we have gotten so far into amperes, then we can complete the balance of our calculation using only amperes.

If you are installing both electric heating and air-conditioning Rule 8-106(4) does not specifically require an interlocking switch. It says simply "where it is known" that both electric heating and air conditioning systems "will not be used simultaneously" we need add only one of these loads, either the electric heating load or the air-conditioning load, whichever is larger. However, some Inspectors may feel that the words "where it is known" means that either a changeover switch must be installed so that only one of these loads can be used at any one time, or that both loads must be added as shown above. The best approach is to add both loads then check if that will bump your service up one size larger. If it does install the larger service or check with your Inspector.

$$\frac{\text{Total watts}}{\text{240 volts}} = ... \text{_____ amps}$$

The following loads are usually added directly in amperes.

STEP 5: ELECTRIC HOT AIR FURNACE & SAUNA

Add 100% of furnace nameplate rating .... =_____ amps
Add sauna at 75% of its nameplate rating .... =_____ amps

STEP 6: AIR CONDITIONING
100% = _____ AMPS. SEE NOTE ABOVE

Total amps .... =_____ .... =_____ amps

Minimum service size is _____ amps

Now consult the List of Materials, p. 5.

EXAMPLE 1

Calculate the service size for a house which has a floor area of 120 m^2 (1291 sq. ft.) on the main floor and 60 m^2 (646 sq. ft.) in the basement, (m^2 x 10.76 = sq. ft.). The electrical load consists of a 14 kw range upstairs and a 12 kw range in the basement, 4 kw. dryer, 3 kw. water heater, 3 kw. sauna and 11 kw electric baseboard heating. There is also a 12 amp A/C unit. Calculate service size.

Basic Load - Floor area = 120 m^2 main floor at 100% = 120 m^2
60 m^2 basement at 75% = 45 m^2

Total = 165 m^2

First 90 m^2 ....5,000 watts
Next 75 m^2 ....1,000 watts

   165 m^2

First range .... 6,000 watts
Plus 40% of 14 kw -12 kw .... 800 watts
2nd range at 25% of rating .... 3,000 watts
Dryer, 25% of 4,000 watts .... 1,000 watts
Water heater, 25% of 3,000 watts .... 750 watts
Electric heating & sauna = 11 kw + 3 kw = 14 kw.

    First 10 kw at 100% .... 10,000 watts
    Balance, 4 kw at 75% .... 3,000 watts

Total =   30,550 watts

$$\frac{30550}{240} = \text{.. 127.3 amps}$$

A/C Unit is 12 amp - Add this at 100% of its rating .... 12 amps

Minimum service size required   139.3 amps

Now refer to the List of Materials, p. 5. For this service use list F.

Our load is 139 amps. This is greater than list C, which is 120 amps, but less than E, which is 150 amps.

Don't ignore the note regarding the panel. The panel for list E may not supply baseboard heaters - but that is what this question requires. Therefore, we must use list F because we need a 40 circuit panel for this load.

SUB-FEEDER SIZES TO SECOND PANEL

It is often an advantage to install a second panel near the kitchen load. The size of panels and feeders and other details are dealt with p. 41.

6.   SERVICE CONDUCTOR TYPES

RULES 12-100 & 12-102

COPPER CONDUCTORS OR ALUMINUM CONDUCTORS

The service conductor may be either copper or aluminum. Aluminum conductor installation requires great care. This special procedure is not covered in this book.

SERVICE CONDUCTOR INSULATION

Service conductors are subject to extreme temperature changes. For this reason the Code requires a different marking on the insulation of service conductors. See the footnotes for Table 19. It requires that we look for surface printed markings on service conductors to show they are certified for use where exposed to the weather. Look for a marking such as "Outdoor use" or similar wording. The old "Minus 40°C" or "-40°C" marking is no longer required although you may still find it on old cable inventory. To my knowledge only the RW90 insulated conductor has the marking required for conductors, which must be formed into drip loops at an entrance cap. If you have difficulty locating properly marked conductors for your service do not just ignore the requirement, consult with your local inspector.

The reason for this restriction is that service conductors, which are connected to overhead lines, are exposed to very low temperatures and to sunlight. Both conditions can destroy the insulation on your service conductors unless they are certified for this application. Accidental contact with a conductor whose insulation has begun to deteriorate could result in severe electrical shock.

This does not apply to the conductors of an underground service because no part of the conductors in an underground service are exposed to the weather.

7.   HYDRO SERVICE WIRES TO THE HOUSE

CONSULT HYDRO

Rules 6-112, 6-116, 6-206, and in Ontario Bulletin 6-1-13: Before any work is done the power utility should be consulted to determine which pole the service will be from. This is very important. The entrance cap must be properly located with respect to the Hydro pole.

There are a number of details to watch out for when locating the service entrance cap.

ENTRANCE CAP ABOVE LINE INSULATOR

Rule 6-116(b) requires the entrance cap to be located between 6 in. and 12 in. above the line insulator. This is to prevent water from migrating along the service conductors to the meter base. See p. 16 for details on this requirement.

ROOF CROSSING

Rule 12-312 - Fire fighting operations require ready access to and free movement on roofs. Check with your Electrical Inspector before crossing any more than the overhang of the roof of any building.

HEAVY SNOWFALL AREAS

The two locations shown along the side of the roof, in the illustration below, may not be acceptable in heavy snowfall areas. Check with your local Inspector.

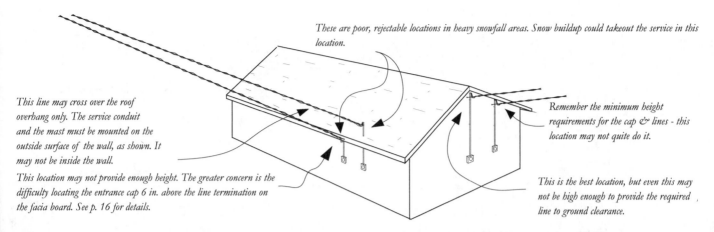

These are poor, rejectable locations in heavy snowfall areas. Snow buildup could takeout the service in this location.

This line may cross over the roof overhang only. The service conduit and the mast must be mounted on the outside surface of the wall, as shown. It may not be inside the wall.

Remember the minimum height requirements for the cap & lines - this location may not quite do it.

This location may not provide enough height. The greater concern is the difficulty locating the entrance cap 6 in. above the line termination on the facia board. See p. 16 for details.

This is the best location, but even this may not be high enough to provide the required line to ground clearance.

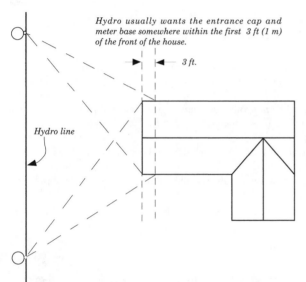

Hydro usually wants the entrance cap and meter base somewhere within the first 3 ft (1 m) of the front of the house.

3 ft.

Hydro line

Hydro usually wants the service entrance cap and meter base to be somewhere in that first 10 ft., 9 in. of the building facing the line, as shown. If it just cannot be located in this part of the house you should check first with Hydro before locating it further back. Cost is usually an important factor but if money is no object I suppose the service could be located almost anywhere on the house.

The front of the house could be the back of the house if the power lines are located in the lane. It refers to the side or end of the house which faces the power lines.

MINIMUM LINE TO GROUND CLEARANCES

Rule 6-112(2) - The line insulator, shown below is required by Hydro for their service drop. It must be installed high enough to provide the following minimum clearances:

ON BC	Meters			Feet		
	Multi-Prov	BC	ONT	Multi-Prov	BC	ONT
Over any public roadway	5.5		6.0	18.0		19.7
Any private roadway or driveway and any other space (including lawns) accessible to commercial and farm vehicles						
Across a readily accessible roof, Rule 12-310		2.5			8.2	
Across residential driveways	4.0		4.5	13.1		14.8
Across walkways, ground accessible to pedestrians only		3.5			11.5	
Sun deck, (this is the same as a flat roof) Rule 12-310 & 315		2.5			8.2	

The line clearances required look like this:

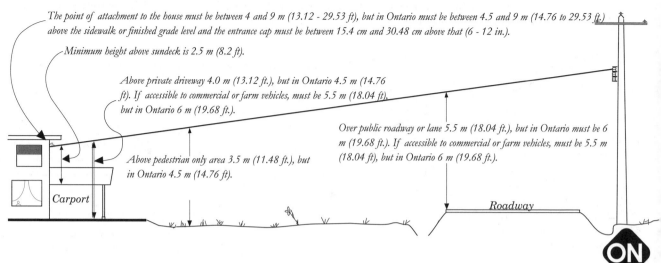

The point of attachment to the house must be between 4 and 9 m (13.12 - 29.53 ft), but in Ontario must be between 4.5 and 9 m (14.76 to 29.53 ft) above the sidewalk or finished grade level and the entrance cap must be between 15.4 cm and 30.48 cm above that (6 - 12 in.).

Minimum height above sundeck is 2.5 m (8.2 ft).

Above private driveway 4.0 m (13.12 ft.), but in Ontario 4.5 m (14.76 ft). If accessible to commercial or farm vehicles, must be 5.5 m (18.04 ft), but in Ontario 6 m (19.68 ft.).

Above pedestrian only area 3.5 m (11.48 ft.), but in Ontario 4.5 m (14.76 ft).

Over public roadway or lane 5.5 m (18.04 ft.), but in Ontario must be 6 m (19.68 ft.). If accessible to commercial or farm vehicles, must be 5.5 m (18.04 ft), but in Ontario 6 m (19.68 ft.).

Carport

Roadway

HYDRO LINE ATTACHMENT TO HOUSE

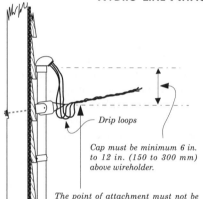

Drip loops

Cap must be minimum 6 in. to 12 in. (150 to 300 mm) above wireholder.

The point of attachment must not be less than 4.5 m (14.8 ft.) nor more than 9 m (29.5 ft.) above the sidewalk or final grade level.

Rule 2-108, 6-112, and 6-116 - Hydro requires that you provide and install an insulator on your building for their line crew to attach their service cable. This wireholder must:

•   Be insulated type, even for triplex cable.

•   Be located within 24 in. of the entrance cap so that proper drip loops can be formed with 75 cm (29.5 in.) leads you must leave hanging out of the entrance cap, Rule 6-302(3). See the illustration above.

•   Be high enough to maintain all the line to ground clearances given in the illustration above but must not be located lower than 4.5 m (14.8 ft.) or higher than 29.5 ft (9 m) above the sidewalk or grade level.

•   Be carefully located so that the 1 m (39.4 in.) line to window and door clearances can be maintained. See illustration, p. 17.

•   Be sufficiently well anchored in a structural member of the building to withstand the pull of the lines in a storm.

Rule 6-112(6) requires that the wire holder must be bolt type as shown above. It may not be lag-screw type.

This means, of course, that if a tree falls across the line to your house or if an over height 70 ton truck comes hurtling down the road and hooks onto the Hydro lines to your house you are safe - your wire holder is designed to withstand these kinds of stresses. It is possible the stress will pull out the wall but you can count on that bolt to not let go of the 2 x 4 stud. It's best to choose a wall that could be pulled out without it taking down the whole house with it.

The construction of the wire holder is an important detail. It must be a type which is designed to hold the lines so they cannot fall to the ground even if the porcelain insulation material on the insulator should break.

The wire holder in the adjacent illustration has a pin holding the porcelain insulator in place. This pin will continue to hold the line even if the porcelain should break and fall away. These guys think of everything.

8.   SERVICE ENTRANCE CAP

For overhead services only

CONSULT HYDRO

Rules 6-112, 6-116(a), and 6-206 - Before beginning the installation of your electrical service, be sure to obtain from your local power supply authority the correct location of the pole from which you will receive service. Your service head must be properly located with respect to the Hydro service pole.

TYPE OF CAP TO USE

Rule 6-114 - There are a number of different caps permissible depending on the type of service conduit or cable you use. See under specific type.

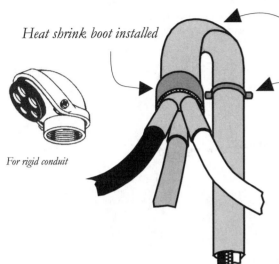

Heat shrink boot installed

The radius of this bend must not be less than six times the external diameter of the TECK cable.

Strap required here.

For rigid conduit

For EMT pipe

Rule 6-114(3) says where a heat shrink boot is used instead of an entrance cap, the cable must be bent downward as shown. Note the additional length required to make this bend.

HEIGHT OF ENTRANCE CAP

Rule 6-112 - There are two things to watch for:
- 1st. That you can obtain the minimum line clearances, p. 15.
- 2nd. Rule 6-116(b) says the entrance cap must be higher than the wire holder which supports the Hydro line. This rule says the cap must be between 15 cm and 30 cm (between 6 & 12 in.) above the Hydro line insulator as shown, p. 16. See "Drip loops" below.

If, in your case, it is difficult to locate the cap above the line insulator you should check with your local Inspector first before you ignore this rule.

BACKGROUND INFORMATION

Rule 6-116(b) was originally designed to prevent water travelling down the Hydro service line to the splices at the entrance cap and there, at the splice, find its way into the conductor itself. Water in this location (actually around the conductor strands inside the conductor insulation) would be drawn along by capillary action between the conductor strands and on down to the meter base where it would collect and do all kinds of nasty things.

DRIP LOOPS

Rule 6-302(3) - This Rule requires that you leave at least 75 cm (29.5 in.) of service conductor hanging out of the entrance cap so that drip loops can be formed as shown, p. 16.

This long length may seem wasteful and sometimes is, but that's what the rule requires. The purpose is to provide sufficient length for Hydro crews to form acceptable drip loops that will prevent water following the service conductors into the service conduit and down into the service equipment as described above.

Conductor Insulation - Rule 12-100 requires that service conductors which are exposed to the weather be suitable for such exposure. See also under Service conductor Types, p. 14.

LOCATION OF ENTRANCE CAP

Rule 6-112(3) - This is an "out of reach" rule. The entrance cap must be placed so that all open conductors (conductors not in conduit or in a cable) are above the window. If they are below or alongside the window, or if they run in front of the window, as shown below, they must be at least 1 m (39.37 in.) away from the window. This rule applies to all windows even those which cannot be opened.

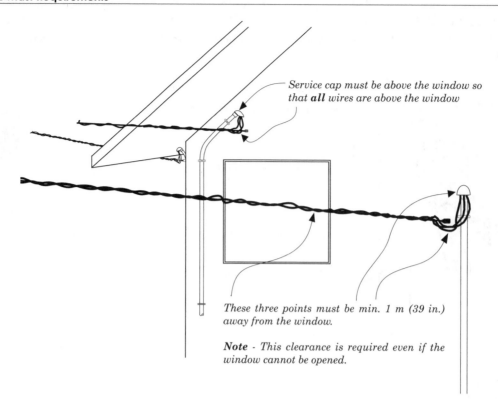

*Service cap must be above the window so that **all** wires are above the window*

These three points must be min. 1 m (39 in.) away from the window.

Note - *This clearance is required even if the window cannot be opened.*

Note 1: A long drip loop is not approved. Locate the entrance cap within 24 in. of the point of attachment of the service drop wires. Hydro service drip leads should contact the building at one point only. Except by special permission these service drop leads may not be run along the building through two or more line insulators to get to the cap location.

Note 2: Snow Slides - Many services have been pulled out or have been severely damaged by snow sliding off a roof during the winter months. It is important that the entrance cap be very carefully located so that such slides cannot harm the service conduit and the Hydro service lines. The gable end of the roof is the preferred location for the entrance cap. The illustrations, p. 15 show a number of possible service locations.

In the on the previous page, one example shows the service leads terminating on the wall below the roof overhang. This is a poor location and may not be acceptable to Hydro because snow sliding off the roof during the winter months could damage these lines — actually pull them out.

Service masts which run through any part of a roof, particularly the lower part of a roof, may also not be acceptable to some utility companies. Therefore, in heavy snowfall districts the entrance cap should, wherever possible, be located on the gable end or similar location on the house where it will not be subject to damage by snow slides.

9.   SERVICE MAST REQUIREMENTS

RULE 6-112(4)(5)(6) & (8)

The illustration below shows the minimum requirements where a service mast is needed to raise the entrance cap and service leads to the required height.

PIPE MAST

These must be mounted on the outside surface of the building, Rules 6-206(1)(e) and 6-208(1).

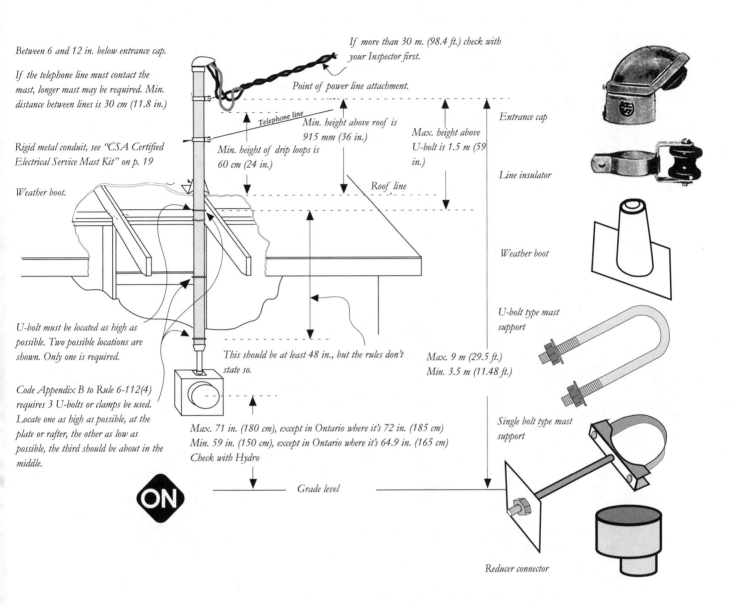

Between 6 and 12 in. below entrance cap.

If the telephone line must contact the mast, longer mast may be required. Min. distance between lines is 30 cm (11.8 in.)

Rigid metal conduit, see "CSA Certified Electrical Service Mast Kit" on p. 19

Weather boot.

U-bolt must be located as high as possible. Two possible locations are shown. Only one is required.

Code Appendix B to Rule 6-112(4) requires 3 U-bolts or clamps be used. Locate one as high as possible, at the plate or rafter, the other as low as possible, the third should be about in the middle.

If more than 30 m. (98.4 ft.) check with your Inspector first.

Point of power line attachment.

Telephone line

Min. height above roof is 915 mm (36 in.)

Min. height of drip loops is 60 cm (24 in.)

Max. height above U-bolt is 1.5 m (59 in.)

Roof line

This should be at least 48 in., but the rules don't state so.

Max. 71 in. (180 cm), except in Ontario where it's 72 in. (185 cm)
Min. 59 in. (150 cm), except in Ontario where it's 64.9 in. (165 cm)
Check with Hydro

Grade level

Entrance cap

Line insulator

Weather boot

U-bolt type mast support

Max. 9 m (29.5 ft.)
Min. 3.5 m (11.48 ft.)

Single bolt type mast support

Reducer connector

If the mast is installed in the path of sliding snow from a sloping roof of smooth hard material such as plastic or metal, you may need to install a guy cable or a brace to support the mast.

PARTS FOR AN ACCEPTABLE MAST

This may appear complicated but all the parts necessary for a service mast can be purchased in most building supply stores. The sales people in these stores will usually assist you in selecting the correct pieces for easy assembly of an acceptable service mast. It is not required that you use a CSA certified service mast kit; masts may be assembled from components suitable for such use.

CSA CERTIFIED ELECTRICAL SERVICE MAST KIT

The rule does not actually say we must use a mast kit, but if you use conduit, its diameter must be 2.5 in. See also under caution below. A mast kit will usually include a length of 2 in. steel pipe (CSA refers to this as 2-3/8 in. but that is its outside diameter), which has been specially tested and certified for this purpose, an entrance cap and a special fitting for the lower end. It will also have the necessary U-bolts or clamps for fastening. Appendix B to this rule requires 3 U-bolts because a mast may extend as much as 59 in. above the roof. There will also be a weather boot and an insulated wireholder complete with pipe clamp so that it can be fastened to the mast as shown above.

If you use a length of standard rigid metal conduit for a service mast Rule 6-112(5) says it must be at least "2-1/2 (63) trade size." Note that the rule does not say "inches" but we can assume that is what is intended. Trade size means nominal size, which is its internal diameter. They are serious about this, we need 2-1/2 in. conduit.

Yes, there are probably a million 2 in. electrical conduit masts out there somewhere serving with valor, but the unamended rule says that "if its conduit 2 in. pipe, it ain't big enough." And yes, it does seem odd that the rule does not require a minimum length below the roof line to support this very strong 2-1/2 in. pipe (the Bulletin requires this but the rule does not), and that it can be bolted to any 2 x 4 in. studded wall. It may also be a little difficult to find the necessary fittings for a 2-1/2 in. pipe and therefore, you may decide to use a mast kit. It is a free country.

WOOD MAST

Wood masts may be acceptable where, as shown in the illustration, a standard metal service mast cannot provide the required clearance for the lines on a flat roof or over a roadway.

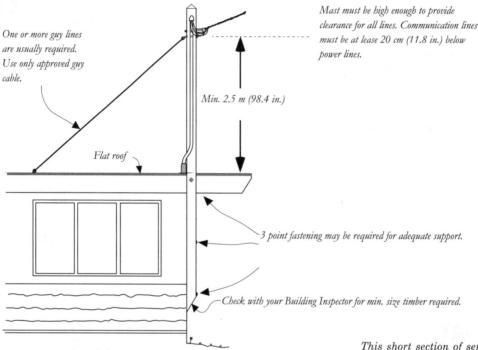

One or more guy lines are usually required. Use only approved guy cable.

Mast must be high enough to provide clearance for all lines. Communication lines must be at lease 20 cm (11.8 in.) below power lines.

Min. 2.5 m (98.4 in.)

Flat roof

3 point fastening may be required for adequate support.

Check with your Building Inspector for min. size timber required.

Some Inspection Authorities have taken the position that a service mast of wood may be regarded as part of the building structure and for that reason is not subject to the Electrical Code.

It is the responsibility of the Building Inspector to determine the size of timber required and the means of its support attachments to the building. Only the location and height of a wood service mast must comply with the Electrical Code as outlined in this book. Check first with your Electrical Inspector before installing a wood mast on any building.

10.   LENGTH OF SERVICE CONDUCTOR

RULE 6-206(1)(E), 6-208

According to these rules the service panel must be located "as close as practicable to the point where the service conductor enters the building." That's a good rule - keep it as short as possible for two good reasons:

(1). Because it is an unprotected conductor - only the Hydro line fuse, which is ahead of the transformer, is protecting you.

(2). The service run is very costly.

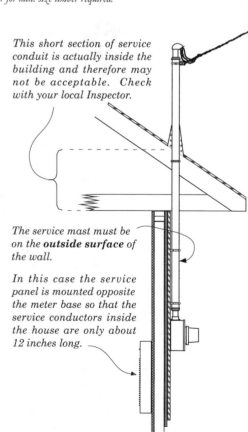

This short section of service conduit is actually inside the building and therefore may not be acceptable. Check with your local Inspector.

The service mast must be on the **outside surface** of the wall.

In this case the service panel is mounted opposite the meter base so that the service conductors inside the house are only about 12 inches long.

MAXIMUM LENGTH PERMITTED OUTSIDE BUILDING

Here the service conductors are in a conduit or in a cable and run on the outside surface of the house the Code does not limit its length. It may be any reasonable length. Cost of material usually keeps this as short as possible.

MAXIMUM LENGTH PERMITTED INSIDE BUILDING

Rule 6-206(1)(e) says the service panel must be "as close as practicable to the point where the consumer's service conductors enter the building." That means just through the wall as shown in the illustration.

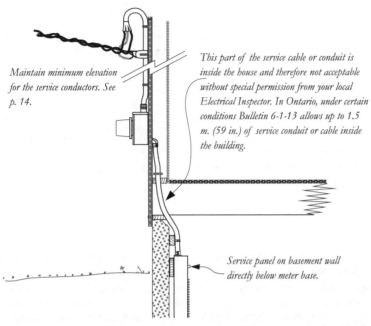

Maintain minimum elevation for the service conductors. See p. 14.

This part of the service cable or conduit is inside the house and therefore not acceptable without special permission from your local Electrical Inspector. In Ontario, under certain conditions Bulletin 6-1-13 allows up to 1.5 m. (59 in.) of service conduit or cable inside the building.

Service panel on basement wall directly below meter base.

The rule says just through the wall but a number of Inspection Authorities have revised the rule with bulletins allowing certain additional lengths. In British Columbia we allow greater, though unspecified, length under certain conditions. In the City of Vancouver they will allow up to 1.5 m (59 in.). In Alberta it is 10 ft. In Saskatchewan it is 6 m (20 ft.). In Ontario, Bulletin 6-1-13 says that under certain conditions it will allow service conductors to run up to 59 in. (1.5 m) into rooms or areas of combustible (wood) construction. In certain other, much more stringent conditions, it will allow as much as 24.6 ft. (7.5 m) of service conductor length into a house.

A clear, decisive, answer is not possible. The best advice is keep it as short as possible, just through the wall, but if you must have more length, explain the situation to your local Inspector and ask for their approval.

The illustrations show a number of situations where short lengths of service conduit or cable enters the house. One of these may help you explain your situation to your Inspector in a request for their approval.

Where concrete encasement is required, it must be at least two inches thick all around the conduit or cable. Where the conduit or cable runs inside a wall check carefully your wall thickness to ensure there will be adequate space to provide for minimum all around covering as shown in the illustrations below.

TECK service cable with PVC jacket may be direct buried in the ground, but thin wall raceway (EMT) must not be.

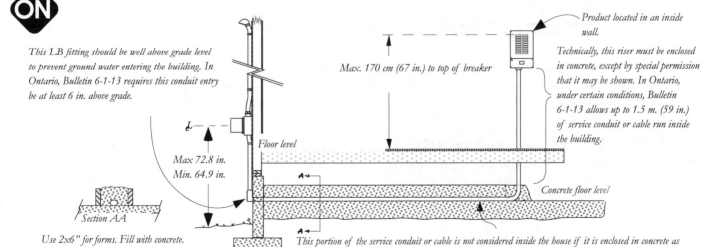

This LB fitting should be well above grade level to prevent ground water entering the building. In Ontario, Bulletin 6-1-13 requires this conduit entry be at least 6 in. above grade.

Max. 170 cm (67 in.) to top of breaker

Product located in an inside wall.

Technically, this riser must be enclosed in concrete, except by special permission that it may be shown. In Ontario, under certain conditions, Bulletin 6-1-13 allows up to 1.5 m. (59 in.) of service conduit or cable run inside the building.

Floor level

Max 72.8 in. Min. 64.9 in.

Concrete floor level

Section AA

Use 2×6" for forms. Fill with concrete.

This portion of the service conduit or cable is not considered inside the house if it is enclosed in concrete as shown, or is in or buried under the concrete floor.

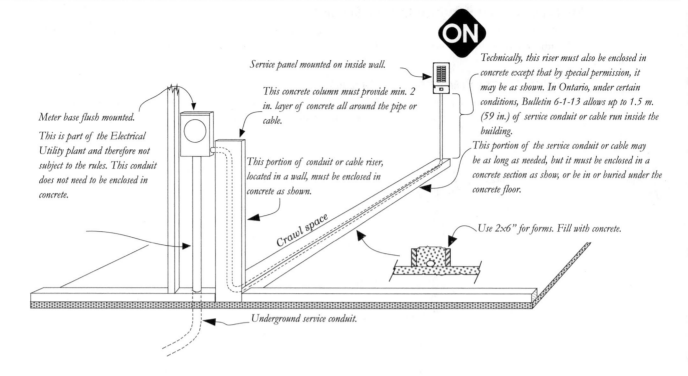

Meter base flush mounted.

This is part of the Electrical Utility plant and therefore not subject to the rules. This conduit does not need to be enclosed in concrete.

Service panel mounted on inside wall.

This concrete column must provide min. 2 in. layer of concrete all around the pipe or cable.

This portion of conduit or cable riser, located in a wall, must be enclosed in concrete as shown.

Technically, this riser must also be enclosed in concrete except that by special permission, it may be as shown. In Ontario, under certain conditions, Bulletin 6-1-13 allows up to 1.5 m. (59 in.) of service conduit or cable run inside the building.

This portion of the service conduit or cable may be as long as needed, but it must be enclosed in a concrete section as show, or be in or buried under the concrete floor.

Crawl space

Use 2x6" for forms. Fill with concrete.

Underground service conduit.

It would be wonderful to have a clear workable standard all across Canada, a maximum length beyond which we could not go without first having to ask for special permission.

Who's to blame? - For more than 40 years we were permitted to run up to 20 ft. of service conduit or cable inside a building in British Columbia then suddenly, with the adoption of the 1994 Code in late 1995, this length was reduced from twenty ft. to less than one foot, just through the outside wall. As an inspector for 30 years I do not know of a single incident where the additional length caused a problem, but I know it solved many problems in finding a suitable location for the service panel. Don't blame your Inspector for this change. It's not their fault.

BOXED IN SERVICE CONDUIT OR CABLE

As noted above, there is no limit to the length of service conduit or cable which is run on the outside surface of the building being supplied. The condition here is that the service conduit or cable is, in fact, on the exterior; run on the outer surface of the building. Some think service conduit or cable is not a thing of beauty, that it should be covered, boxed in, that it is beautiful only when out of sight.

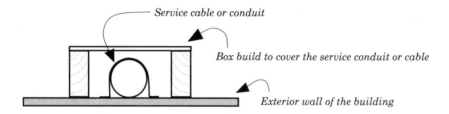

Service cable or conduit

Box build to cover the service conduit or cable

Exterior wall of the building

Before boxing-in exposed service conduit or cable, to hide it from view, you should check with your Inspector. The rule says service conductors must be as short as possible after they "enter the building." Since such boxing may be considered part of the building, the service conduit or cable, therefore, enters the building where it enters the boxing. Such an installation may therefore not be acceptable unless the authority enforcing the Code has agreed to accept it.

11.   SERVICE CONDUIT OR SERVICE CABLE

There are several different wiring methods permitted for service conductors:
- EMT thin wall conduit method,
- Rigid PVC Conduit method,
- Cable, such as TECK 90.

EMT - Thin Wall Service Conduit

Electrical metallic tubing (EMT) thinwall metal tubing used to not be acceptable for an electrical service, but some provinces now permit it for exposed and concealed work, wet and outdoor locations, and in or on buildings of combustible or non-combustible construction (Rule 12-1402). Because service conduit is frequently exposed to wet conditions, EMT may still not be acceptable for service work in Ontario. Check with your local Inspector. The remainder of this section from this page and running up to "Rigid PVC Conduit" on p. 25 is not applicable in Ontario, but is applicable to all of the other provinces that have adopted the CSA Electrical Code.

Rule 6-302(1)(c) This is permitted by Code but in some regions close to salt water, where corrosion is a problem, it is not acceptable for services. Check with your local Electrical Inspector.

This is a thin wall conduit which cannot be threaded. It requires very few tools for installation.

Length

See p. 20 for maximum length permitted inside the building.

Couplings & Connectors

Rule 12-1410: Set Screw Type and Compression Type - There are two main types of thinwall raceway conduit couplings and connectors used today.

Compression Coupling   Set Screw Coupling   Compression Connector   Set Screw Connector

Screw Type -These may be used indoors or directly under a roof overhang or in concrete but may not be exposed to the weather.

Compression Type may be used in all cases wherever EMT is permitted.

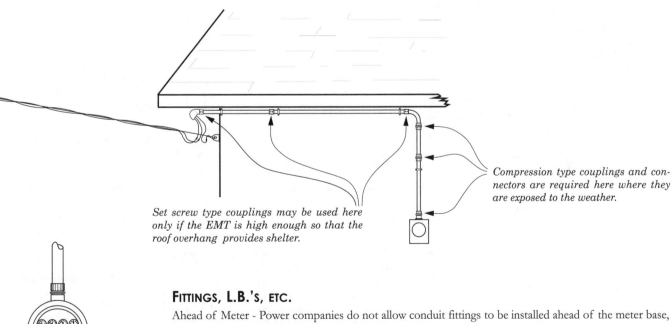

Set screw type couplings may be used here only if the EMT is high enough so that the roof overhang provides shelter.

Compression type couplings and connectors are required here where they are exposed to the weather.

L.B. Fitting

Fittings, L.B.'s, etc.

Ahead of Meter - Power companies do not allow conduit fittings to be installed ahead of the meter base, however, on the load side of the base you may install as many as you require. Where it cannot be avoided and a fitting must be installed ahead of the base, you should seek Hydro permission first before proceeding.

Problems using L.B. fittings - Avoid using these fittings where service conductors are larger than #3 copper. Great care is needed when seating, or forming conductors into the fitting. Pull them into place one at a time being careful not to damage the insulation. It's okay to use a hammer to form the conductors into the fitting but do not apply directly on the insulation. Hold a smooth piece of wood against the conductors and drive that gently with a hammer.

Double L.B. Problem - Never use two L.B. fittings back to back for any conductors larger than #6. It is very difficult to force the conductors into the fitting without damaging the insulation in the process. Replace one of the fittings with a manufactured 90° bend.

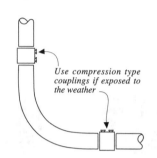

Avoid these conduit fittings if at all possible even on the load side of the meter because it can be very difficult to get the conductors to lie properly in the fitting. The conductors for 100 amp and larger services are very stiff.

Never use two fittings back to back as shown when the conductors are #4 or larger.

Use compression type EMT connectors where they are exposed to the weather.

It is too difficult to force these conductors into this second LB fitting. Replace this second LB fitting with a quarter bend.

Use compression type couplings if exposed to the weather

Accessible - L.B. Fittings must be Accessible - These must be located where they will remain accessible for any maintenance work that may be required in the future.

BENDS

Rule 12-1108 & 12-1112: Manufactured Bends - The maximum permitted is the equivalent of four quarter bends (4 - 90° bends). These must be made without damage to the pipe. You can bend this pipe yourself but not on the truck bumper or around a tree. Bends must be made with a bending tool, called a hickey. You should be able to rent one from the local tool rental shop. The pipe must remain round, not oval in shape and there should not be any kinks and no, you may not heat this pipe (EMT), to bend it.

SERVICE (METAL) CONDUIT BONDING

Rule 10-604 - Locknuts and bushings are no longer acceptable for bonding EMT service raceway. service conduit bonding. A grounding type bushing and a jumper, as shown below, must be used and these must be installed where shown.

Size of jumpers - Use a # 8 copper jumper for any size service up to 100 amp and a #6 copper jumper for a 101 to 200 ampere service, Code Table 41.

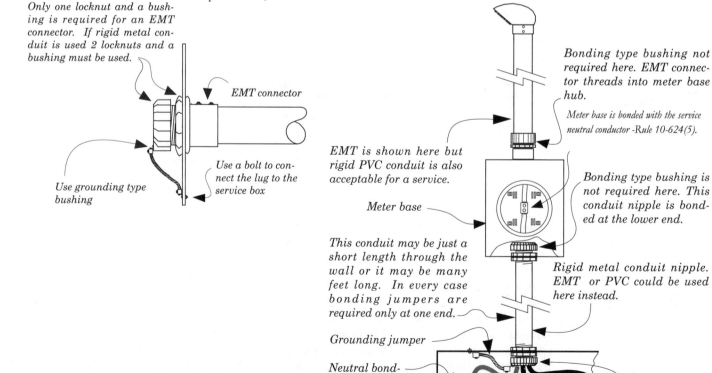

Only one locknut and a bushing is required for an EMT connector. If rigid metal conduit is used 2 locknuts and a bushing must be used.

EMT connector

Use a bolt to connect the lug to the service box

Use grounding type bushing

Bonding type bushing not required here. EMT connector threads into meter base hub.

Meter base is bonded with the service neutral conductor -Rule 10-624(5).

EMT is shown here but rigid PVC conduit is also acceptable for a service.

Meter base

Bonding type bushing is not required here. This conduit nipple is bonded at the lower end.

This conduit may be just a short length through the wall or it may be many feet long. In every case bonding jumpers are required only at one end.

Rigid metal conduit nipple. EMT or PVC could be used here instead.

Grounding jumper

Neutral bonding screw

Service grounding conductor

Grounding type bushing

This bonding jumper is in addition to the locknuts which connect the conduit at both ends. It may be a bit much, we haven't needed it for forty years, but now we do - it's the law.

Note (a) Rule 10-624(4) Says that where the neutral runs through the metal meter base, (it almost always does) the meter base must be bonded to ground with the service neutral conductor as shown above.

Note (b) If the nipple between the meter base and the panel is rigid metal conduit, not EMT, then two locknuts are required in addition to the bonding jumper shown above.

Note (c) Locknuts are dished, that is they are not flat. They are designed to bite through the paint or rust and into the metal of the box. This is necessary to provide good grounding for the equipment. Make sure you install them with the sharp edges facing the panel wall where they will cut through the paint when they are tightened.

Note (d) Bushings must be grounding type and where the conductors are #8 or larger, the bushings must have an insulating ring, such as plastic or nylon, mounted in the throat to protect the service conductors. See above for an illustration of such a bushing.

STRAPPING

Rules 12-1114 - Install one strap within 1 m (39 in.) of the top, (entrance cap) end, another within 1 m (39 in.) of the meter base. Install additional straps between as required so that the maximum distance between straps is not more than 2 m (approx. 6 ft.).

SEALING

Rule 6-312 - The seal required by this rule prevents the warmer inside air from escaping through the service conduit. It has been found that the warm air, if allowed to flow, condenses to water in sufficient quantity to damage the service equipment.

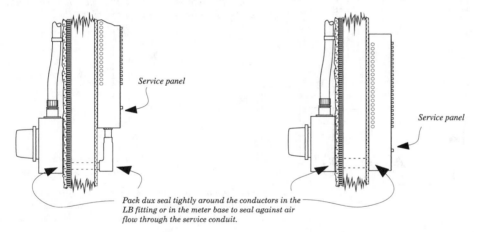

Service panel

Service panel

Pack dux seal tightly around the conductors in the LB fitting or in the meter base to seal against air flow through the service conduit.

This seal is usually made with a soft putty-like substance called DUX Seal. To be most effective the DUX Seal is placed around the conductors in the last opening, before the conduit leaves the warmer area.

RIGID PVC CONDUIT

Rule 6-302(1)(a) - In most cases PVC conduit is acceptable for services. It has an advantage - very few tools are required to install it and there is no need for bonding bushings and bonding jumpers. Make sure you do the following:

• Sun Exposure - If your PVC service conduit will, at any time, be exposed to direct rays of the sun, then it must be approved and marked for such exposure. Look for an SR or "Sunlight resistant" marking.

• Rough Burrs - Be sure to remove rough burrs from the inside edge of the pipe ends, where it has been cut. Do this with any knife.

• Clean and Dry - Be sure the pipe and fittings are clean and dry before applying solvent cement.

The illustration shows the arrangement of fittings required when installing PVC service conduit. The female adapter and close metal conduit nipple shown above the meter base are required to prevent this conduit breaking at this point, Rule 12-1112(2). The PVC terminal adapter used below the meter base is acceptable only where it enters an enclosure through a knock-out hole; it may not be used in a threaded hole such as the top entry into the meter base.

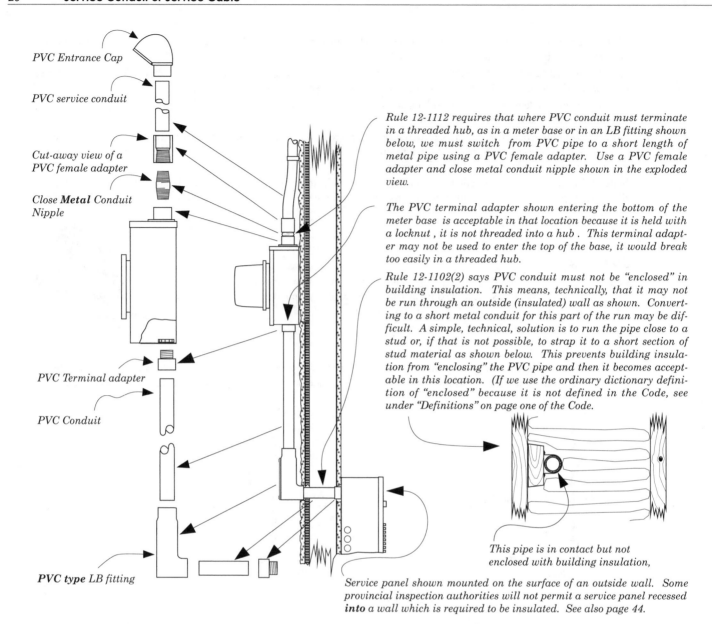

PVC Entrance Cap

PVC service conduit

Cut-away view of a PVC female adapter

*Close **Metal** Conduit Nipple*

PVC Terminal adapter

PVC Conduit

***PVC type** LB fitting*

Rule 12-1112 requires that where PVC conduit must terminate in a threaded hub, as in a meter base or in an LB fitting shown below, we must switch from PVC pipe to a short length of metal pipe using a PVC female adapter. Use a PVC female adapter and close metal conduit nipple shown in the exploded view.

The PVC terminal adapter shown entering the bottom of the meter base is acceptable in that location because it is held with a locknut, it is not threaded into a hub. This terminal adapter may not be used to enter the top of the base, it would break too easily in a threaded hub.

Rule 12-1102(2) says PVC conduit must not be "enclosed" in building insulation. This means, technically, that it may not be run through an outside (insulated) wall as shown. Converting to a short metal conduit for this part of the run may be difficult. A simple, technical, solution is to run the pipe close to a stud or, if that is not possible, to strap it to a short section of stud material as shown below. This prevents building insulation from "enclosing" the PVC pipe and then it becomes acceptable in this location. (If we use the ordinary dictionary definition of "enclosed" because it is not defined in the Code, see under "Definitions" on page one of the Code.)

This pipe is in contact but not enclosed with building insulation,

*Service panel shown mounted on the surface of an outside wall. Some provincial inspection authorities will not permit a service panel recessed **into** a wall which is required to be insulated. See also page 44.*

Fittings - L.Bs, L.Ls, etc.

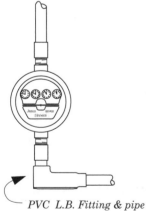

Power companies do not normally allow conduit fittings to be installed ahead of the meter base, however, on the load side of the base you may install as many fittings as you require. Where it cannot be avoided and a fitting must be installed ahead of the base, you should seek Hydro permission first before proceeding.

Problems using L.B. fittings - Avoid using these fittings where service conductors are larger than #4 copper. Great care is needed when seating, or forming conductors into the fitting. Pull them into place one at a time being careful not to damage the insulation. It's okay to use a hammer to form the conductors into the fitting but do not apply directly on the insulation. Hold a smooth piece of wood against the conductors and drive that gently with a hammer.

Double L.B. Problem - Never use two L.B. fittings back to back for any conductors larger than #4. It is very difficult to force the conductors into the fitting without damaging the insulation in the process. Replace one of the fittings with a 90° bend.

PVC L.B. Fitting & pipe

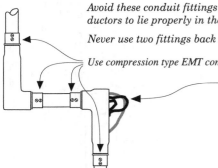

Avoid these conduit fittings if at all possible even on the load side of the meter because it can be very difficult to get the conductors to lie properly in the fitting. The conductors for 100 amp and larger services are very stiff.

Never use two fittings back to back as shown when the conductors are #4 or larger.

Use compression type EMT connectors where they are exposed to the weather.

It is too difficult to force these conductors into this second LB fitting. Replace this second LB fitting with a quarter bend.

L.B. Fittings must be Accessible - These must be located where they will remain accessible for any maintenance work that may be required in the future.

BENDS

Rules 12-1108 - Manufactured Bends - The maximum number of bends permitted by the rule is the equivalent of four quarter bends (4 - 90° bends or any combination of bends where the total does not exceed 360°), Rule 12-942.

HOME MADE BENDS

Not Recommended - Avoid this if you can, but if your installation requires a special bend you can take advantage of the fact that PVC conduit may be bent in the field. To do this the pipe must be carefully heated to 260° F at the location of the proposed bend. It is best to use a heat gun to heat the pipe. An open flame could also be used, the Code does not prohibit it, but in that case it must be done very carefully to avoid damaging the pipe. Too much heat will char the pipe or blister its smooth surface and that could be grounds for rejection. Make sure the bend section is uniformly heated all around for a distance of about 10 times the pipe diameter before you attempt to bend it. Improper heating or improper bending procedures may cause the pipe to collapse, or even worse, kink and that would very likely result in a rejection of your installation. One more thing, there is a small amount of spring back when the pipe cools. To compensate for this loss you will need to over bend it a few degrees more than is required.

PVC pipe & couplings

STRAPPING

Rule 12-1114 - Install one strap at the top, within 1 m (39 in.) of the entrance cap, another strap near the bottom, within 1 m (39 in.) of the meter base. Install additional straps as required so that the maximum distance between any two straps is not more than 2 m (approx 6 ft.).

Note that PVC conduit has a high coefficient of expansion. This means that it is a bit longer in the summer than in the winter. If it is exposed to direct sunlight, it is even longer. For short runs, 10 ft. or less, this is not a great concern, but longer runs should be supported with straps that permit the conduit to slide when it expands with temperature change.

SEALING

Rule 6-312 - The seal required by this rule prevents the warmer inside air from escaping through the service conduit. It has been found that warm air, if allowed to flow through the service conduit, will condense to water in sufficient quantity to damage service equipment.

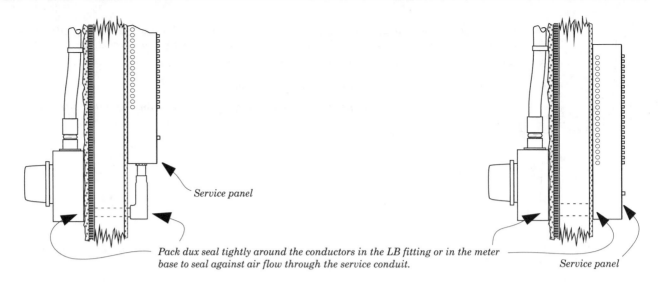

Service panel

Pack dux seal tightly around the conductors in the LB fitting or in the meter
base to seal against air flow through the service conduit.

Service panel

The seal is usually made with a soft putty like substance called DUX Seal. To be most effective DUX Seal is carefully placed to close all openings around the conductors in the last opening before the conduit leaves the inside of the house or in the first opening outside of the house. Only one seal is required. Make sure all openings around each conductor are completely sealed to prevent any air passage through the conduit.

HOLES IN OUTER WALLS, FLOORS OR ROOFS

Rules 12-018 & 12-928 - These rules require that we fill in any openings around conduit or cable where these pass through an outer wall or a roof or a floor. Rule 12-928 requires that where your service conduit enters your building from an underground electrical distribution system the end of the conduit must be sealed to prevent gas or moisture from entering your house. Depending on the grade, this may require a substantial seal to prevent water entering through the electrical service conduit. Check with Hydro for advice on the type of seal necessary for your service conduit. This is not exactly electrical work, but your Electrical Inspector will check this part of your installation and they are required to yell at you, or stomp their foot if this is not properly done for the final inspection.

SERVICE CABLE - FROM AN OVERHEAD HYDRO SUPPLY

TYPE

Rule 6-302(1) says types ACWU75, ACWU90, AC90 or TECK-90 cables may be used.

Alcan multi-conductor TECK-90 cable

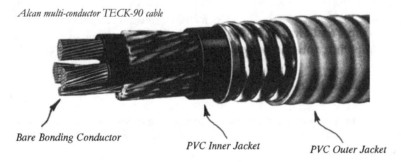

Bare Bonding Conductor

PVC Inner Jacket

PVC Outer Jacket

If you plan to install service cable where, at any time, it will be exposed to the direct rays of the sun then it must be approved and marked with an SR or Sunlight Resistant marking, Rule 2-134.

Service cable is easier to install than conduit. Each of these service cables differs from the others in some way. The illustration shows the construction of a TECK-90 cable.

SIZE

The ampacity required for a service cable is the same as for an equivalent size conductor in conduit. See p. 5 for sizes required.

TYPE OF ENTRANCE CAP TO USE

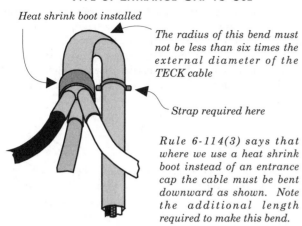

Heat shrink boot installed

The radius of this bend must not be less than six times the external diameter of the TECK cable

Strap required here

Rule 6-114(3) says that where we use a heat shrink boot instead of an entrance cap the cable must be bent downward as shown. Note the additional length required to make this bend.

Rule 6-114 requires that when using armoured cable such as TECK cable for the service it must be equipped with an acceptable entrance cap. Entrance caps designed for use with conduit should not be used on cable. If you use a metal entrance cap it must be bonded to ground with the bonding conductor in the cable. Run the bare wire through one of the holes in the cap and then terminate it in a connector lug on the base of the metal entrance cap To avoid these problems try one of the following suggestions.

Entrance caps designed for EMT are not certified for use with service cable. They are equipped with set screws which hold the cap in place on EMT, but if used on a cable could damage the cable insulation. They are economical, fit quite well, and are often used on cable, but check first with your Inspector before installing an EMT type entrance cap on a TECK cable. Don't forget rule 10-300 requires the metal entrance cap to be bonded to ground.

Heat Shrink Entrance cap - Rule 6-114(2) & (3). This is a kind of plastic sock that slips over the top end of your service cable. To install this sock you must remove the outer PVC jacket, the armour and the inner PVC jacket for approx 30 in. You should now have 30 in. of insulated conductors exposed for the drip loops and for connection to Hydro lines.

Slip this sock in place, pull it on top of the PVC jacket on the cable, then very carefully apply heat, as evenly as possible, with an open gas torch or heat gun, (heat lamps could also be used) as long as the heat applied is approximately 250°F. Apply heat evenly and do not overheat the material. When heat is applied the sock will shrink to fit snugly in place. It will seal it from rain but

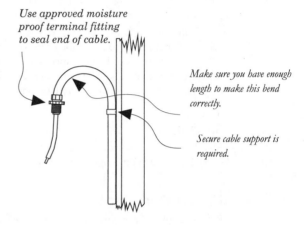

Use approved moisture proof terminal fitting to seal end of cable.

Make sure you have enough length to make this bend correctly.

Secure cable support is required.

remember, the rule says it must still face downward when installed. Be sure to allow enough length to do this. The bonding conductor in the cable at the entrance cap is simply cut off.

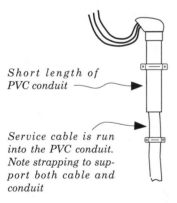

Short length of PVC conduit

Service cable is run into the PVC conduit. Note strapping to support both cable and conduit

A Weatherproof Cable Connector - Rule 6-114(2&(3) - The entrance cap in this illustration is a weatherproof cable connector. The problem here is the same as in the upper illustration, the cable must be long enough to face down as shown. The radius in this cable bend must be at least 6 times the cable diameter.

Bonding for this metal connector is provided by the correct installation of the connector. Follow the instructions provided by the manufacturer of the connector.

A PVC Conduit Entrance Cap - Use a PVC entrance cap and a short length (12 in.) of PVC conduit. Bond the cap to the conduit with PVC cement and strap it in place as shown.

This arrangement should be confirmed with your Inspector before proceeding.

The bonding conductor in the cable at the entrance cap, is simply cut off.

CABLE CONNECTORS

Use weatherproof connectors where the cable connects to the top of the meter base. Be sure to use the correct size and type for the cable you are using. Your supplier will advise you on this.

Dry type connectors may be used only where they are not exposed to the weather.

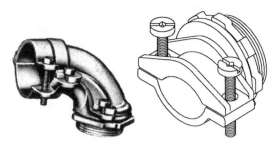

Anti-Short Bushings, shown adjacent, must be used with dry type connectors. This is a fibre or plastic bushing that fits into the end, inside the armour, of the cable. It protects the conductor insulation at the point where they issue from the armour. This thing is wholly inside the connector but its presence can be easily verified through the small openings provided by the manufacturer for this detective work. It is easier to put it in place when the cable is being installed. Later, after it has been rejected, it is much more difficult to do this.

LENGTH OF SERVICE CABLE PERMITTED

Rule 6-206 & 6-208 - Outside of the house - Service cable, which is run along the outside surface of the house, may be as long as it needs to be to get where you're going. However, this cable is expensive; keep it as short as possible.

Anti-short bushing

Service cable run inside the house - The rule says service equipment, (the service panel) must be located so that the length of service cable run inside the house can be kept as short as practicable. See p. 20 for details on maximum lengths and the exceptions permitted.

Very Long Service Runs - Rule 6-208 - Where the panel location is well within the building we can take advantage of another rule which permits us to install long runs of service cable inside a building if it is enclosed in at least 2 in. of solid concrete (all around covering) or if it is buried in the ground under the floor. The illustrations, p. 20 show such an installation for conduit. A cable would be installed a little differently but the principle remains the same; any conduit or cable which is enclosed in at least 2 in. of concrete is considered to be outside the house and may, therefore, be as long as needed.

STRAPPING

Rule 12-618 - Cable must be strapped within 30 cm (12 in.) of its termination point at each end and every 1.5 m (59 in.) throughout the run.

NEUTRAL CONDUCTOR

Rule 4-028(4) - The neutral conductor in the cable must be white or grey. If the cable does not have a white or grey conductor it's okay to paint the neutral white or grey. Make sure that you paint the same conductor at each end where the outer covering has been removed to make connections. This identification is critical.

METER CONNECTIONS

Rules 10-516(2) & 10-906(3)

Do not cut away the PVC jacket on this cable, it must run into the connector *in this wet location.*

The bare bonding wire must terminate in a separate grounding lug which is bolted to the side wall of the meter box, as shown. Do not use one of the wood screws for this purpose, Rule 10-906(3).

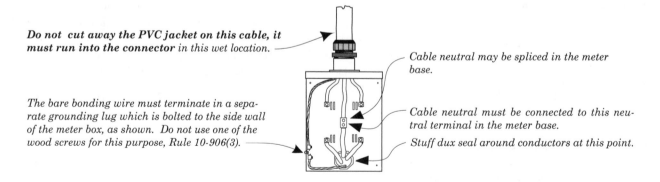

Cable neutral may be spliced in the meter base.

Cable neutral must be connected to this neutral terminal in the meter base.

Stuff dux seal around conductors at this point.

MECHANICAL DAMAGE

Greater care is needed when installing this cable than is required when installing conduit. This cable is more easily damaged with driven nails or where it is run on the surface of a wall.

Where the cable is run on the surface of the wall and where it may be subject to mechanical damage (in locations such as a garage or carport for example) the cable may be protected with wooden or metal guards or a short section of metal pipe may be used.

In the illustration below, the cable is shown running through the plate and over the broken edge of the foundation wall. This section of cable (where it runs through the plate) is subject to damage and must be protected. Use heavy gauge metal plates to protect the cable at these points. The side plates of metal sectional outlet boxes do this very well.

The illustration below shows TECK cable being used above the meter base to an overhead supply and from the meter base to the panel. It also shows PVC conduit from the meter base downward to an underground Hydro system. Obviously, only one of these systems will be available - either overhead supply or underground supply and therefore only the TECK cable running up or the PVC conduit running down is necessary. Both systems are shown to indicate different methods of installation.

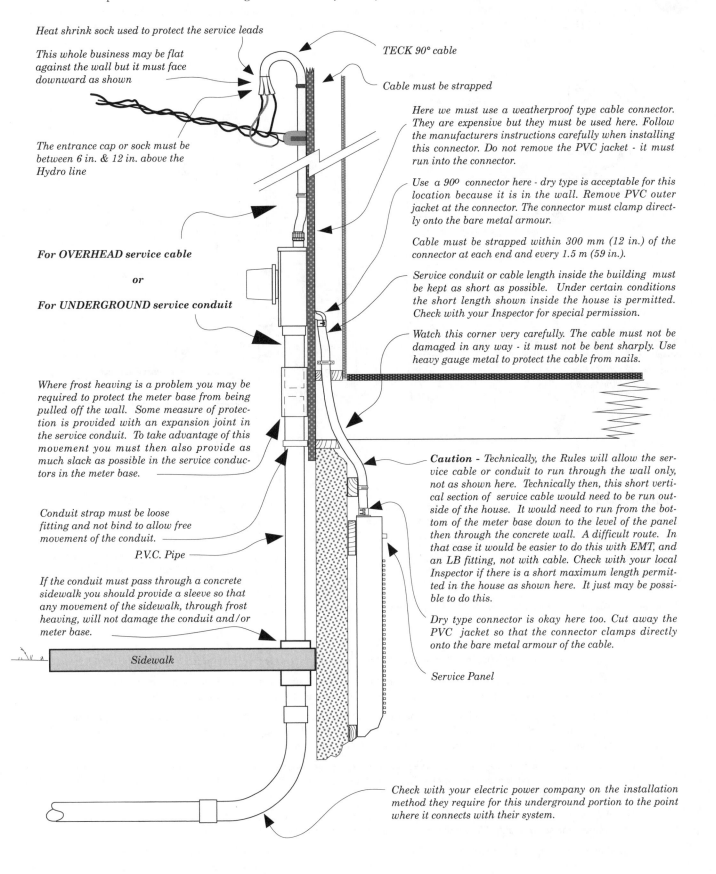

Heat shrink sock used to protect the service leads

This whole business may be flat against the wall but it must face downward as shown

The entrance cap or sock must be between 6 in. & 12 in. above the Hydro line

For OVERHEAD service cable

or

For UNDERGROUND service conduit

Where frost heaving is a problem you may be required to protect the meter base from being pulled off the wall. Some measure of protection is provided with an expansion joint in the service conduit. To take advantage of this movement you must then also provide as much slack as possible in the service conductors in the meter base.

Conduit strap must be loose fitting and not bind to allow free movement of the conduit.

P.V.C. Pipe

If the conduit must pass through a concrete sidewalk you should provide a sleeve so that any movement of the sidewalk, through frost heaving, will not damage the conduit and/or meter base.

Sidewalk

TECK 90° cable

Cable must be strapped

Here we must use a weatherproof type cable connector. They are expensive but they must be used here. Follow the manufacturers instructions carefully when installing this connector. Do not remove the PVC jacket - it must run into the connector.

Use a 90° connector here - dry type is acceptable for this location because it is in the wall. Remove PVC outer jacket at the connector. The connector must clamp directly onto the bare metal armour.

Cable must be strapped within 300 mm (12 in.) of the connector at each end and every 1.5 m (59 in.).

Service conduit or cable length inside the building must be kept as short as possible. Under certain conditions the short length shown inside the house is permitted. Check with your Inspector for special permission.

Watch this corner very carefully. The cable must not be damaged in any way - it must not be bent sharply. Use heavy gauge metal to protect the cable from nails.

Caution - Technically, the Rules will allow the service cable or conduit to run through the wall only, not as shown here. Technically then, this short vertical section of service cable would need to be run outside of the house. It would need to run from the bottom of the meter base down to the level of the panel then through the concrete wall. A difficult route. In that case it would be easier to do this with EMT, and an LB fitting, not with cable. Check with your local Inspector if there is a short maximum length permitted in the house as shown here. It just may be possible to do this.

Dry type connector is okay here too. Cut away the PVC jacket so that the connector clamps directly onto the bare metal armour of the cable.

Service Panel

Check with your electric power company on the installation method they require for this underground portion to the point where it connects with their system.

12. SERVICE FROM AN UNDERGROUND SUPPLY

HYDRO CONNECTION

Your local Hydro authority will have certain requirements for an underground service installation. To begin your inquiry with local Hydro staff you may want to show them the illustration on the prior page for a ducted system or the illustration below for a direct buried service cable.

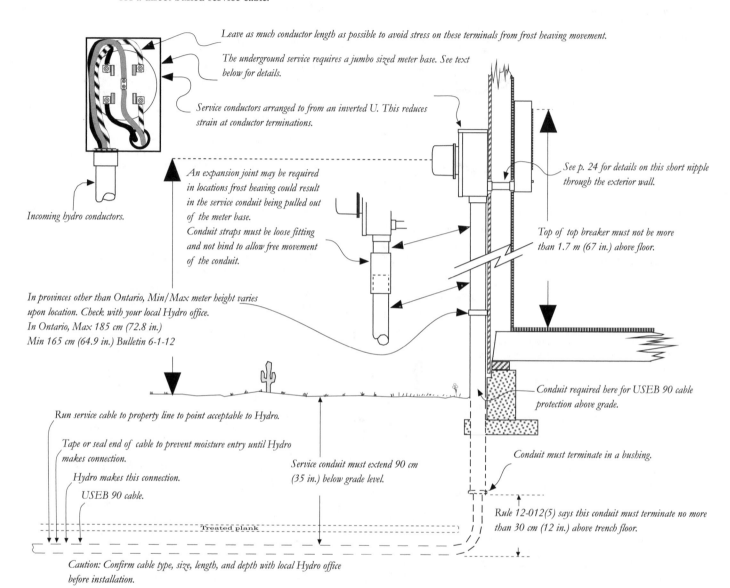

Leave as much conductor length as possible to avoid stress on these terminals from frost heaving movement.

The underground service requires a jumbo sized meter base. See text below for details.

Service conductors arranged to from an inverted U. This reduces strain at conductor terminations.

An expansion joint may be required in locations frost heaving could result in the service conduit being pulled out of the meter base.
Conduit straps must be loose fitting and not bind to allow free movement of the conduit.

Incoming hydro conductors.

In provinces other than Ontario, Min/Max meter height varies upon location. Check with your local Hydro office.
In Ontario, Max 185 cm (72.8 in.)
Min 165 cm (64.9 in.) Bulletin 6-1-12

See p. 24 for details on this short nipple through the exterior wall.

Top of top breaker must not be more than 1.7 m (67 in.) above floor.

Conduit required here for USEB 90 cable protection above grade.

Run service cable to property line to point acceptable to Hydro.

Tape or seal end of cable to prevent moisture entry until Hydro makes connection.

Hydro makes this connection.

USEB 90 cable.

Service conduit must extend 90 cm (35 in.) below grade level.

Conduit must terminate in a bushing.

Rule 12-012(5) says this conduit must terminate no more than 30 cm (12 in.) above trench floor.

Treated plank

Caution: Confirm cable type, size, length, and depth with local Hydro office before installation.

Before proceeding with your installation you need to obtain the following information from local Hydro staff:

(1). Which of the two arrangements is acceptable to them? - The illustration provides an opportunity for Hydro staff to know what you plan to install and to make whatever revisions necessary to comply with their regulations.

(2). What is the exact location and depth of the trench required for the cable? - Normally it is the owner's responsibility to prepare the trench. Check with Hydro for the exact location of their underground tap at the property line. Check with Hydro regarding depth of burial, type and size of cable and who must provide it.

(3). What type and size of underground service cable is required for your installation? - In order for Hydro to determine the size of the underground service cable they will need to know the size of your service. Determine the size of your service from the Table, p. 10

(4). Who supplies and who installs and connects the underground service cable? - Most likely Hydro will supply, install and connect the service cable. They will advise you about this procedure.

(5). Are there any special size requirements for the meter base - In some cases Hydro may require a 200 Amp meter base for a 100 amp underground service. One of the concerns is the space needed for sufficient additional conductor length to allow for changes in length due to frost heaving.

(6). Service Cable Protection Above Ground - The Electrical Code requires protection for the service cable where it rises above ground to enter the meter base. The Code is satisfied with any size of pipe provided the cable can pass through easily but your power company may have a word or two to say about the size of this conduit. Do check out this detail and remember that your meter base must be deep enough to accept whatever pipe size is required.

A few more items to be aware of:

• Sand Bed - Both Hydro and the Electrical Code require a 75 mm (3 in.) layer of sand both below and above the service cable.

• Do not cover underground cable until authorized by the Electrical Inspector.

• The connector lugs - Make certain that all the connector lugs inside the meter base and the service box are properly rated for the size of underground service conductors you plan to install.

• Peace of Mind - Check with your local power company to make sure there are no other hiccups that could cost you time and money when you finally ask them for that power connection.

Length of Service Run Inside the House - This must be kept as short as possible which means only the length required to run through the exterior wall of the house. See p. 20 for detailed information.

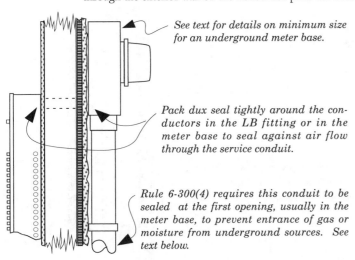

See text for details on minimum size for an underground meter base.

Pack dux seal tightly around the conductors in the LB fitting or in the meter base to seal against air flow through the service conduit.

Rule 6-300(4) requires this conduit to be sealed at the first opening, usually in the meter base, to prevent entrance of gas or moisture from underground sources. See text below.

Sealing - Rule 6-312 - The seal required by this rule prevents the warmer inside air from escaping through the service conduit. It has been found that the warm air, if allowed to flow, condenses to water in sufficient quantity to damage the service equipment.

This seal is usually made with a soft, putty-like substance called DUX Seal. To be most effective the DUX Seal is placed around the conductors in the last opening before the conduit leaves the warmer area.

Drainage of Underground Ducting - Pay careful attention to gradients of underground service ducting. Where your service duct is lower than the Hydro ducting at the property line you will need seals and drainage holes to be drilled into the service duct at specific locations to prevent water entering service equipment. Check with Hydro for their requirements.

Hydro Connection - Contact Hydro for location of trench termination at the property line.

SERVICE FEEDER TO HOUSE

For the underground portion from the property line to the house there are at least two choices as far as the rules are concerned. These are as follows: Conductors may be installed in conduit as shown in the illustration on p. 31, or as shown below, service cable certified for direct burial without conduit may be used. Check with your local Hydro office if they have a preference.

SIZE OF SERVICE CONDUIT

Building Code Rule 9.34.4.5 - The Building Code requires that this vertical section of conduit from the trench to the meter base must be at least 2 in. in diameter. However, bear in mind that each of the separate power utilities in Ontario may set its own requirements for this service conduit. They may require this conduit to be larger than 2 in. but it may not be smaller than 2 in.

DEPTH OF BURIAL

Electrical Code Rule 12-012(5) - End of Service Conduit - The Building Code requires this service conduit to extend at least 90 cm (35 in.) into the trench and Rule 12-012(5) requires this conduit to terminate 30 cm (12 in.) above the trench floor. That straight section of cable from the end of the service conduit to the trench floor is provided to permit movement as a result of frost heaving.

Depth of Service Cable - Check with your local Hydro office for minimum depth of service cable burial. Each utility will have its own standards. Some utilities will provide and install this cable in a trench you have provided others may not provide or install it. In some cases you will be required to install Big O pipe for the service cable. This pipe is not acceptable as a service conduit, it serves only as a chase for quick removal of the service cable in the event of trouble or upgrade. For this reason, when service cable is drawn into Big O pipe it is considered to be direct buried, therefore, the cable must be suitable for direct burial. Cable such as USEB is acceptable for this service.

Caution tape is required when running underground cable.

BASE

For a service supplied from an Underground Distribution System - The Bulletins do not specify a 200 Amp meter base for every service supplied from an underground system nor does it require a specific sized meter base when installed for an underground service. However, meter base size is important. It must have a deep enough enclosure for the underground service conduit to enter through the bottom wall.

SEALING

Rule 6-312 - The seal required by this rule prevents the warmer inside air from escaping through the service conduit. It has been found that the warm air, if allowed to flow, condenses to water in sufficient quantity to do serious damage the service equipment.

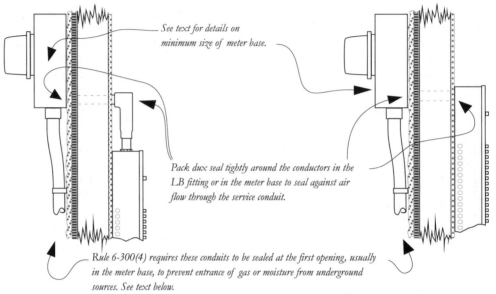

See text for details on minimum size of meter base.

Pack dux seal tightly around the conductors in the LB fitting or in the meter base to seal against air flow through the service conduit.

Rule 6-300(4) requires these conduits to be sealed at the first opening, usually in the meter base, to prevent entrance of gas or moisture from underground sources. See text below.

This seal is usually made with a soft, putty-like substance called DUX Seal. To be most effective the DUX Seal is placed around the conductors in the last opening before the conduit leaves the warmer area.

Underground Service Conduit from an Underground Distribution System - Rule 6-300(4) requires service conduit to be sealed at the first opening above ground. This is to prevent stray gas or moisture entering the enclosure, (the meter base). Dux seal used to prevent breathing through the conduit nipple which runs through the wall can also be used here to seal against gas entering the meter base. Where the meter base is below the level of an underground service piping system from the property line Dux seal is not a sufficient barrier to prevent water entry. Check with Hydro office for advice on an acceptable seal.

Conduit Entry into a House from an Underground Distribution System - Rule 6-300(4) requires the underground service conduit to enter a house above finished grade. It may not enter directly below ground. There is a good reason for this requirement. Gas leaking from a nearby underground gas line could travel along the electrical pipe and enter the house. The volume of gas entering would likely not amount to much but if it became trapped and began to accumulate it would only need a tiny spark to cause a violent explosion. Where the conduit emerges outside of the house explosive gas vapors can evaporate safely to the atmosphere.

13. METER BASE INSTALLATION

TYPES

In general there are two types of meter bases available, the round and the square or shoe box type. Both types are acceptable for services supplied from overhead lines. Make sure your base has the correct rating in amperes.

Note 1: For an underground service the power utility usually requires a 200 ampere meter base for all underground services up to 200 amp rating. This large base may be necessary to provide space for additional conductor length to allow for frost heaving movement. It may also be necessary to provide sufficient depth for the conduit entry. This is illustrated on p. 32.

Note 2: If the knockouts provided by the manufacturer of the base are not in the correct position and new holes need to be punched out, these new holes must be totally below any live parts in the meter base. Only where a meter base is installed totally indoors, as in a service room in an apartment building, may new holes be cut above the live parts in the base.

Old Round Base - By the way, if you plan to use an old round type meter base which has side or back conduit entry holes, be careful. First, because some utilities do not accept them and second, only the top and bottom entries in round bases may be used where these round bases are allowed at all. One more thing, many of those old meter bases were not designed to permit splicing of the neutral in the base. This means the round base may not be suitable if you are using TECK cable for service conductors. The round base is not acceptable for an underground service.

LOCATIONS

Bulletin 6-1-13 and Ontario Building Code Rule 9.34.4.2(2) - Power companies usually want the meter base as close to the front of the house as possible. The illustration on p. 14 shows that the preferred location of the service entrance cap and meter base The meter base must be located within the first 3 ft (118 in.) of the front of the house. The front of the house is the side nearest the Utility distribution system.

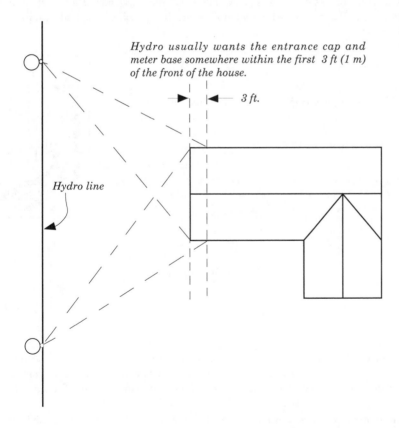

Hydro usually wants the entrance cap and meter base somewhere within the first 3 ft (1 m) of the front of the house.

3 ft.

Hydro line

Meter locations must be carefully chosen. Some of the things to watch for are:

(1). Carports - If you plan to face your meter base into the carport you should know that there is no code rule or bulletin which specifically says you may not have it there. However, before you install it in the carport you should talk to your local Hydro people. They may not like it there.

Fact is, no matter how careful you are as a driver, when you are backing in your 50 ft. Winnebago that meter is subject to damage if it faces into the carport.

Every year there are thousands of carports closed in to make a garage or an additional bedroom. If the meter faces into the carport the space cannot be closed in to convert it into a bedroom without first relocating the meter base. This is usually a costly relocation.

(2). Porch - If it is an open porch it may be an acceptable location now but remember you may want to close it in later on in the future. A closed in porch is a heat saver in the winter time, so avoid the hassle, follow old Chinese proverb - don't do it on the porch.

Ventilation Rule 2-320 and Bulletin 2-10-6 in Ontario - This Code now takes a more relaxed approach when evaluating a possible hazardous location. The concern is not with the gas meter itself but with the relief vent which is usually located near the gas meter. It is the discharge point of this vent that presents a possible hazard. The space around a gas meter vent would not become hazardous except as a result of an accident, rupture or breakdown. This means it is a Class 1, Zone 2 location, Rule 18-006(c)(1). However, Rule 2-320 still requires 1 m (39.4 in.) separation between a gas relief vent on a gas meter and an electric meter or any other possible spark producing electrical device such as a light or plug outlet.

HEIGHT

The height of the meter base is a very simple detail yet it is not the same in every province. For provinces other than Ontario, they are normally located somewhere between 150 cm (59 in.) and 180 cm (71 in.) above finished grade. In Ontario, they are located somewhere between 165 cm (64.9 in.) and 185 cm (72.8 in.) above finished grade. This is measured from grade level to the center of the base. Please confirm this at your local power company office.

CONNECTIONS

Rule 10-624(4) requires the neutral to be connected to the meter base as shown. All modern meter bases have provision for connecting the neutral from both the line and the load. Some of the older meter bases may not be properly equipped to make a splice in the neutral conductor. In those cases simply bare a section of the conductor where it passes the bonding terminal in the meter base - then slip it into the lug provided and tighten. Do not cut this conductor unless it is necessary to do so.

The hot line conductors, and the hot load conductors (These are usually the black and red conductors) should be connected as shown below. Rule 10-204 (and Bulletin 10-15-1 in Ontario) require the neutral service conductor to be connected to the meter base as illustrated below.

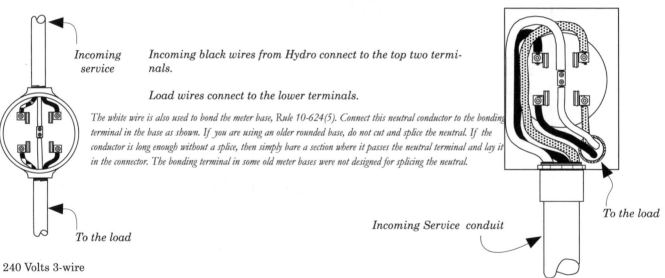

Incoming service

Incoming black wires from Hydro connect to the top two terminals.

Load wires connect to the lower terminals.

The white wire is also used to bond the meter base, Rule 10-624(5). Connect this neutral conductor to the bonding terminal in the base as shown. If you are using an older rounded base, do not cut and splice the neutral. If the conductor is long enough without a splice, then simply bare a section where it passes the neutral terminal and lay it in the connector. The bonding terminal in some old meter bases were not designed for splicing the neutral.

To the load

Incoming Service conduit

To the load

240 Volts 3-wire

Bonding of Meter Base - As noted above the neutral conductor in a service for a single family house must always be connected to the meter base as shown above even in those cases where the base is already directly connected with metal conduit to the grounded metal service panel . The neutral connection is critical in the event of a short circuit in the base and in every case where non-metallic conduit is used between the meter base and the service panel.

This is also a serious concern in a dip service, which this book does not cover, where the service conductors from the meter base to the panel in the house may be in a non-metallic conduit, so that the neutral connection in the base is the only grounding means for the metal base. Local Rules may also require a #6 copper bonding conductor to run with the service conductors from the service panel to the meter base.

SUPPORT

Support the meter base with wood screws through the two or more factory drilled holes in the back of the base. Make sure it is in a reasonably accurate upright position.

BLANK COVER

In some districts the power utility will energize the electrical service but not install a meter until several days later. During this time a blank cover is required to prevent anyone coming in contact with live parts. A disk of 1/4 in. plywood may also be used for this purpose.

SEALING RINGS

Meter sockets are equipped with either a screw type sealing ring or a spring type ring. Some Power Company's will not accept the spring clip type ring. Check with your power company.

FITTINGS & L.Bs ETC.

Ahead of Meter - Some power companies do not allow fittings to be installed ahead of a meter base. This has something to do with stealing power which power company's have decided is bad for business. However, on the load side of the base you may install as many as you require. Where it cannot be avoided and a fitting must be installed ahead of the base you should seek Hydro permission before proceeding. They may be satisfied if the fitting is within sight of the meter and the fitting has been drilled to permit Hydro personnel to install a seal. See also the notes on p. 23 and p. 26 regarding two L.B. fittings.

14. SERVICE PANEL

TYPE - FUSE OR BREAKER PANEL

Both are acceptable, however, only the circuit breaker panel is in common use today. For this reason we will deal with circuit breaker panels only. The illustration below shows a typical breaker panel and the cut-away in the cover shows the connections required for 3 wire cables.

L.B. Fitting

3-Wire Cables & Tie-Bars

When Are These Required - Note the different uses for 3-wire cables.

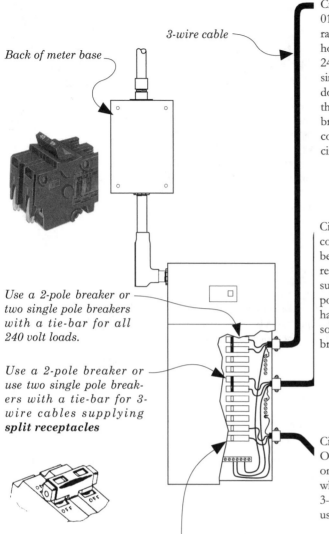

3-wire cable

Back of meter base

Use a 2-pole breaker or two single pole breakers with a tie-bar for all 240 volt loads.

Use a 2-pole breaker or use two single pole breakers with a tie-bar for 3-wire cables supplying **split receptacles**

*Tie-bar not required but **hot wires must connect to adjacent breakers** as shown.*

Circuit Breaker for 3-Wire Cables Supplying 240 Volt Loads - Rule 14-010(b) is concerned with safety for anyone working on equipment such as ranges, dryers, or electric heaters which are supplied with 240 volts. Two hot wires, and sometimes a neutral, are required in the supply cable to serve 240 volt equipment and both of the hot lines in this cable must be opened simultaneously whenever the supply breaker is turned off. The best way to do this is with a two-pole circuit breaker which will open both hot lines with the operation of one handle. However, the rule also permits two single pole breakers to be used instead, provided the operating handles are mechanically connected together with an approved tie-bar so that with one operation both circuit breakers are opened.

Circuit Breaker For 3-Wire Cables supplying split duplex receptacles -Kitchen counter plug outlets are required to be split duplex receptacles and these must be supplied with 3-wire cable. These outlets supply only 120 volts but each receptacle is in fact supplied with 240 volts because both hot lines in the 3-wire supply cable are connected to it. For this reason the rule requires either a two-pole breaker or two single-pole breakers (with a tie-bar to link their operating handles) to supply these 3-wire cables. Note that Rule 26-700(11) now requires some kitchen counter plugs to be GFCI protected and in such cases the supply breaker must be a 2-pole 15 amp GFCI type. See p. 85 for details.

Circuit Breaker for 3-Wire Cables supplying lights and Ordinary Duplex Plug Outlets - Lights and convenience plug outlets (except split duplex receptacles on the kitchen counter) are connected to only one hot line and the neutral when supplied with a 3-wire cable. Circuit breakers used to protect these 3-wire cables need not be 2-pole type nor do we need to install tie-bars when using two single pole breakers to supply these loads.

When connecting 3-wire cables in the panel make very sure that the black and the red wires are connected to two different circuit breakers which are located side-by-side, (one above the other). The reason for this is not to install a tie-bar because, as noted above, a tie-bar is not always required by the rules. The reason is load balance, and thus safety. The neutral conductor in a correctly connected 3-wire cable carries only the unbalance current, usually much less than either the black or the red wire. However, if the 3-wire cable is incorrectly connected the neutral wire must carry the sum of the loads in the black wire and the red wire and this could cause overheating of this neutral wire. Make sure each 3-wire cable is connected to breakers located next to each other, as shown in the panel above.

Bathroom Razor Outlet

Rule 26-700(11) - These must be supplied with a GFCI, (Ground Fault Circuit Interrupter) type circuit breaker or you may use a ground fault circuit interrupter type receptacle. The old special transformer type razor outlet is no longer approved and may not be installed in new construction but may only be installed to replace a faulty unit in an existing installation. A better alternative is to replace the old transformer type plug outlet with a special faceplate that fits over the large existing outlet box and transforms it so that a single GFCI type receptacle may be installed. Note that in some older installations there may not be a bonding conductor in the bathroom outlet. Rule 26-700(8) now allows a GFCI type receptacle to be used even when there is no bonding conductor in the outlet box; however, in general, GFCI type receptacles are considered supplementary protection but not a substitute for proper insulation and grounding.

IDENTIFY CIRCUITS

Kinds of Loads Served - Rule 2-100(2) & (3) (and Bulletin 2-5-3 in Ontario) - Use a felt pen or some other permanent manner of marking next to the circuit breaker or fill in the circuit directory card provided by the panel manufacturer. The identification should look something like this.

LR, Lights and plugs, 15 Amps
Fam/den Lights and plugs, 15 Amps
Lights and plugs, 15 Amps
Outdoor plugs, 15 Amps
Outdoor plugs, 15 Amps
Carport plug, 15 Amps
Dryer outlet, 30 Amps
Spare circuit for future use
Kitchen counter plugs, 20 Amps
Kitchen counter plugs, 20 Amps
Kitchen counter plugs, 20 Amps
Compactor, 15 Amps
Instant hot water, 15 Amps
Sauna heater, 30 Amps
Etc.

Bsmt Lights and plugs, 15 Amps
BR Lights and plugs, 15 Amps
Range outlet, 40 Amps
Lights and plugs, 15 Amps
Lights and plugs, 15 Amps
Fridge outlet, 15 Amps
Laundry outlet, 15 Amps
Garbage disposal unit, 15 Amps
Furnace, 15 Amps
Spare circuit for future use
Kitchen counter plugs, 15 Amps
Kitchen counter plugs, 15 Amps
Etc

BASEMENT OR ATTIC SUITE PANEL

Rule 26-400 & 26-722(a) - In an Existing House - Rule 26-400 was changed in the previous Code to make it less threatening and less costly to comply with. The rule says that where an existing single family house is being renovated to create one or more rental suites, all outlets in the one or more suites may be served from one panel.

Check the existing service size to ensure it is adequate to supply the new suite load. In most cases, if your service is 100 amperes and heating is not with electricity, service size is very likely sufficient for the additional load. If there is any doubt about service ampacity calculate the total load as follows:

(1). Calculate the load in the existing dwelling (not the new suite load) as shown on p. 12; then

(2). Calculate the load in the new suite as if it were another separate house, see p. 12 (Note In this calculation for steps 1 & 2, leave out all air-conditioning and electric heating loads); then

(3). Add 100% of the largest calculated load to 65% of the smaller load; then

(4). Add all air conditioning loads and/or the electric heating load, whichever is greater, as shown on p. 12; then.

(5). Determine the size of service required to serve the total load of both suites using the table on p. 10 and the List of Materials on p. 5.

If the ampacity of the existing service is not adequate, (according to the above calculations) to serve the existing house load and the new suite load, then the existing service must be upgraded to supply the additional load, or another separate, (metered) service could be installed to serve only the new suite load. See "Caution" below.

If the ampacity of the existing service is adequate for the new suite load but there is insufficient space in the panel for the additional breakers that are needed for the new circuits, the panel only must be replaced with a larger one, or a second panel must be installed.

Location of this second panel - It is not required to be located in the new suite, (it may be located next to the existing service equipment and supplied from it) nor is it required that all the circuits for the new suite be supplied from this second panel. Each of the circuits from either suite may be supplied from either panel, as convenient. If, however, your future plans call for separate metering for each suite you should separate the loads now in this renovation when it is easier to do that.

Subfeeder to Second Panel may be supplied with a set of breakers in the existing service. To provide space for the two Subfeeder breakers you may need to reroute two circuits from the old panel to the new.

For the Student - Rule 26-400(2) applies only when Subrule (1) requires a separate panel but if we are creating a suite in an existing single family dwelling Subrule (1) does not require a separate panel, therefore, Subrule (2) does not apply in the above example.

Size of Subfeeder cable to this second panel may be determined as shown in the above calculation. Remember, this is not just a second panel in a single occupancy; it serves a self contained suite with cooking and branch circuit loads. It may also have its own heating and hot water tank. Make sure the Subfeeder is large enough to carry whatever load is connected to this panel.

By way of explanation - Subrule 26-722(a) should also have been revised to bring it more completely into line with the revision to Rule 26-400. These two rules could easily be seen to be in conflict. The latest revision to Subrule 26-400 was made, primarily, to ease the pain for owners creating a suite in the basement. It was intended to remove the penalty for declaring an illegal suite. Before this latest change these two rules required a separate panel in each suite and all circuits to be redirected so that the panel in each suite would supply only the outlets in that suite. It was an extremely expensive requirement.

Owners did not want to declare a suite because they feared the authorities would require expensive alterations be made. As a result many suites may be unsafe because they were created without any inspection of the work. That fear was removed by the revision of Rule 26-400. We can now come clean, fess up and ease our consciences. Go ahead, say it, "I have a suite in my basement!"

If the separate panel noted above is separately metered, then all the full wrath of the rules will be applied. It is not that Hydro dislikes the extra meter, but that you now have, in effect, a separately metered duplex or triplex and all the expensive electrical rules and building rules for such buildings must then be applied.

In a New House - The major concession in the rules refers only to an existing single family dwelling which is being renovated to provide a rental unit in the basement or elsewhere. It does not apply to a new house which is constructed to include a self contained suite. In new houses each suite (if there are more than one) must be served with its own branch circuit panel which is located in the suite which it serves. In that case all the circuits in both suites must also comply with Rule 26-722(a). This subrule says branch circuits may serve only the outlets in the suite where the panel is located; there may not be any mixing of any of the loads. All this applies in a new house under construction.

Separate metering for each panel is not required by these rules but often is very desirable. Consider locating the suite panel so that service changes for separate metering can be made in the future with the least possible difficulty.

MINIMUM CIRCUITS REQUIRED

The number of circuits needed for a given house is determined by the minimum service ampacity as shown on the Service Ampacity Table, p. 10 —AND— the number of outlets installed in that house.

The table on p. 10 has been designed to simplify this problem. This table specifies the required ampere rating of the service and indicates which list of materials, given on p. 5, should be used for that particular house. Each list of materials also indicates the size of panel required.

There are two steps involved - proceed as follows:

(1). Determine minimum number of circuits required from the Table, p. 10.

The table on p. 10 gives the minimum service ampere rating required. Next to it is a letter in brackets. This letter refers to the list of service material required, p. 5. This list also indicates the minimum number of circuits required for that house.

(2). Complete the chart, p. 41 to determine the actual number of circuits needed to supply the outlets you plan to install. Carefully fill in the chart to make sure you do not run short of circuits when the loads are finally being connected. Don't forget the last entry in this table. You must have at least two totally unused circuit spaces in the panel which are not used or required for any load.

Result - The size of the branch circuit panel must be equal to step 1 or step 2, whichever is greater.

Before doing this, see p. 54.

Light Outlets - count all light outlets. Indoor and outdoor,
(do not count switch outlets) .... _____

Convenience Plug Outlets - This refers to:

        Living room plugs .... _____
        Family room plugs .... _____
        Bedrooms plugs .... _____
        Dining room plugs .... _____
        Any Other Rooms or areas .... _____

Each Bathroom   Minimum 1 plug receptacle required .... _____
                 Bathroom fan .... _____

Each Washroom Minimum 1 plug receptacle required .... _____

Each Hallway Minimum 1 plug receptacle required
for each hallway. See p. 78 .... _____

Each Kitchen fan Counts as one outlet See p. 54 and p. 119 .... _____

               Total Outlets Required = ... _____

     Total outlets required

then = _____ = Circuits required .... _____
           12

(12 outlets is the max. load permitted. It is better to divide by 10, or even by 8, for fewer outlets per circuit)

ADDITIONAL CIRCUITS REQUIRED

Outdoor plug outlets   Minimum 1 circuit required, see p. 92.. _____

Carport or Garage   Minimum 1 circuit required in each,
see p. 94 and p. 94 .... _____

Laundry room or area   Minimum 1 circuit required, see p. 91...................................... _____

Kitchen   Minimum 1 circuit for fridge... _____

Plus counter outlets, see p. 82.. _____

Larger Appliances Range - 2 circuits required.. _____

2nd. Range 2 circuits required.. _____

Dryer - 2 circuits required.. _____

Dishwasher - 1 circuit required.. _____

Garburator - 1 circuit required.. _____

Compactor - 1 circuit required... _____

Microwave - 1 circuit required... _____

Instant Hot Water Heater - 1 circuit required................................... _____

Hydro massage Bath-tub - 1 GFCI circuit required....................... _____

Built-in vacuum Cleaner - 1 circuit required.................................. _____

Furnace (gas or oil) - 1 circuit required... _____

Electric furnace - 2 circuits req'd. See p. 124................................. _____

Electric baseboards - See p. 121 to determine circuits req'd. ...... _____

Boiler - 1 circuit required... _____

Domestic Use Water Heater - 2 circuits required............................ _____

Swimming pool - Motor Load. 1 circuit required............................ _____

Lighting load - 1 circuit required .... _____

Freezer plug (Separate circuit not required but is better)..................................... _____

Sauna Heater - 2 circuits required .... _____

Hot tub Motor load - 1 circuit required .... _____

Electric heater - 2 circuits required .... _____

Domestic Water Pump - 1 circuit required (check pump rating)........................... _____

Any Special plugs or lights .... _____

Plus 2 spare circuits (Rule 8-108(2)).. _____

_____

Total Circuits Required = _____

SUB-FEEDER TO SECOND PANEL

Sometimes it is an advantage to install a second panel to supply the electrical loads in specific areas such as a kitchen or a garage. The kitchen requires a number of separate circuits for special loads such as the fridge, microwave, compactor etc. The basement area directly below the kitchen is usually a good location for that second panel. Remember, the code does not require it. All the loads in the house may be served from the main service panel, however, this may require a very large service panel and a lot of costly long home runs. For this reason it may be an advantage to install a second panel.

Don't forget, this second panel must be located with the same care and attention you used to locate the main service panel. All the rules regarding panel location, height, accessibility etc. as outlined below for a main service panel, must also be applied to the second panel. It may not be put just anywhere.

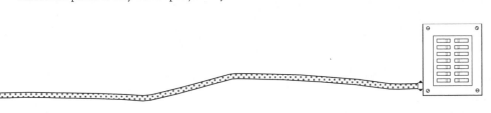

Service panel

Sub-panel

SIZE OF PANEL AND FEEDER CABLE REQUIRED

Section 8 of the code has nothing to say about minimum sizes for a second panel in a single family dwelling. We cannot apply Rule 8-200 to a sub-panel because that rule must be applied to the whole house, not just to a part of the load, and it sets the minimum size at 60 amps. This rule cannot, therefore, be properly applied to our sub-feeder.

However, all is not lost. We may apply a thumb rule to arrive at satisfactory sizes. The "Thumb Rule" goes like this: For any size house up to approximately 4,000 sq. ft. floor area the following is usually acceptable.

EXAMPLE 1

Lighting Loads Only - If the second panel will supply only lighting loads:
- Sub-feeder Size #10 copper loomex cable. 30 amp fuses or circuit breakers in the main panel.
- Sub-panel Size 8 or 12 circuit panel. We may have as many as we wish. This is usually governed by the number of outlets per circuit, the area served and the bank account. Make sure your panel is large enough for the present load and for some future load additions.

EXAMPLE 2

Kitchen Electrical loads - If the second panel supplies lights and plugs in the kitchen area as well as other loads such as the garburator, dishwasher compactor etc. but not an electric range or dryer or any electric heating:
- Sub-feeder Size #8 copper loomex cable. 45 amp fuses or circuit breakers.
- Sub-panel Size 12 or more circuit panel is recommended. It is better to have too many circuit spaces for what you need now than to have too few spaces for your loads now and nothing for future load additions.

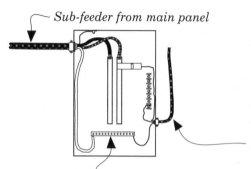

Sub-feeder from main panel

Connect the black & red wires to the buses.

The white wire connects to the neutral bus.

The bare wire in the supply cable connects to the enclosure at each end. This is an important connection. All grounding for all these branch circuits depends on these connections.

Only one branch circuit shown

There should be no connection between the neutral bus and the metal enclosure - if there is a bonding screw, or a bonding jumper in this panel, remove it and throw it away.

Note 1: The range, dryer, and electric heating loads are not normally supplied from the second panel. These loads are usually supplied directly from the main service panel.
Note 2: For a private Garage Panel - see "Private Garage and Farm Buildings," p. 132.
Note 3: Additions & renovations to an existing house - See p. 136 for details.

LOCATION OF SERVICE PANEL

Rule 6-206 (and Bulletins 6-3-0 & 6-4-0 in BC) - The service equipment must be inside the building served.

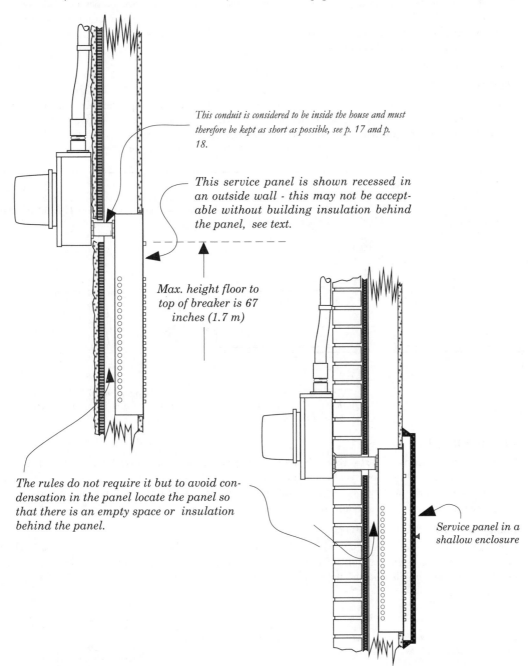

This conduit is considered to be inside the house and must therefore be kept as short as possible, see p. 17 and p. 18.

This service panel is shown recessed in an outside wall - this may not be acceptable without building insulation behind the panel, see text.

Max. height floor to top of breaker is 67 inches (1.7 m)

The rules do not require it but to avoid condensation in the panel locate the panel so that there is an empty space or insulation behind the panel.

Service panel in a shallow enclosure

In previous Codes it was thought necessary to provide a space or a sheet of gyprock behind the panel. This was to prevent overheating of the wood members which are in contact with the back of the panel. In the event of an overheated circuit breaker the back wall of the panel could also become overheated. The space, or the gyprock, is no longer required by Code but it may still be worthwhile to provide for it.

The illustrations show a service panel recessed into an exterior wall. Such an installation is acceptable to Code but could result in a lot of condensation inside the panel unless there is space or building insulation behind the panel. The Electrical Code does not require this space but some Inspection authorities do suggest that there should be at least 12 mm (0.5 in.) of empty space or 12 mm (0.5 in.) of insulation behind the panel. It is best to mount the panel on the surface of an exterior wall. If that is not possible, the next option may be to recess it in a partition. The third option may be to recess it in an exterior wall but with space behind it.

ACCESSIBILITY

The service panel should be mounted in a free wall space where it will remain accessible. It should not be located above freezers, washers, dryers, tubs, counter space, etc. It should not be on the back wall of a storage room where access may become difficult due to stored items, nor should it be in bathrooms, clothes closets, stairwells, kitchen cabinets or similar places, Rules 6-206, 2-308(1) & 26-402.

Where such equipment is flush mounted in a wall, a covering door may be installed over the equipment for appearance sake. This cabinet would be very shallow, providing no storage space for other items. Sometimes a calender or large picture is hung over the panel. This always provides a certain amount of excitement when a fuse blows, all the lights go out and you cant remember whether the panel is under that old picture of uncle George, who knew nothing about electricity, or the Mona Lisa, who seems to think it funny you cant find the panel.

HEIGHT

Rules 6-206(1)(b) & 26-402(2) - Minimum height above floor - The code no longer specifies a minimum height for the panel. The revised rule now simply says the panel must always be placed as high as possible but never more than 1.7 m (67 in.) to the top of the top breaker in the panel. Aim for the maximum height, 67 in. above the floor. A few in. less may be acceptable, but if it is any more than about five in. lower you should check with your inspector first.

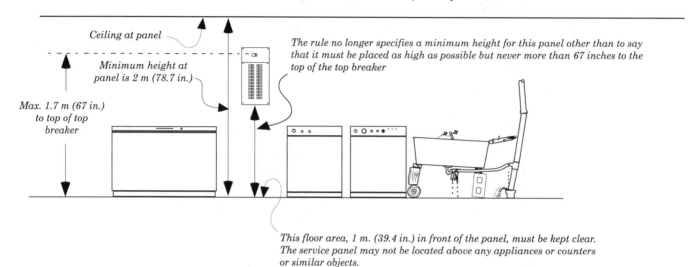

Ceiling at panel

Minimum height at panel is 2 m (78.7 in.)

Max. 1.7 m (67 in.) to top of top breaker

The rule no longer specifies a minimum height for this panel other than to say that it must be placed as high as possible but never more than 67 inches to the top of the top breaker

This floor area, 1 m. (39.4 in.) in front of the panel, must be kept clear. The service panel may not be located above any appliances or counters or similar objects.

WHICH END IS UP?

Service circuit breaker panel boards should be mounted in a vertical position although there is no rule that actually says so, (circuit breakers will function in either the horizontal or vertical position). There is, however, the question of, of, well, professionalism. Panels mounted in a vertical position do look more handsome, don't you think?

Most fused service switches may not be mounted upside down or on their side, Rule 14-502.

ARRANGEMENT OF CONDUCTORS

Rule 6-212, 12-3034 - Inside your service panel is a barrier that divides the space into two separate sections. The main service breaker is (usually) in the top section and the branch circuit breakers are (normally) in the lower section although there is nothing wrong with installing it with the main section at the bottom.

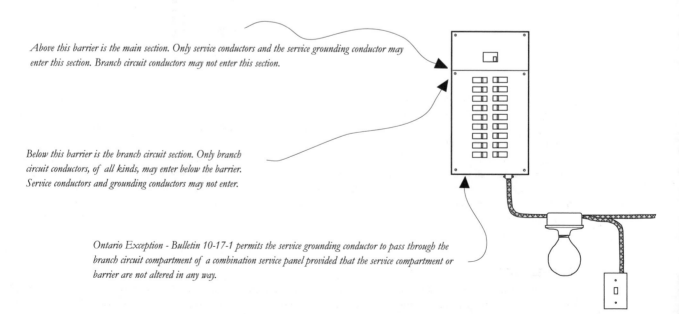

Above this barrier is the main section. Only service conductors and the service grounding conductor may enter this section. Branch circuit conductors may not enter this section.

Below this barrier is the branch circuit section. Only branch circuit conductors, of all kinds, may enter below the barrier. Service conductors and grounding conductors may not enter.

Ontario Exception - Bulletin 10-17-1 permits the service grounding conductor to pass through the branch circuit compartment of a combination service panel provided that the service compartment or barrier are not altered in any way.

SERVICE SECTION

May contain only the service conductors and the service ground. Branch circuit conductors may not enter, or run through, this section.

BRANCH CIRCUIT SECTION

May contain only branch circuit conductors of all kinds. Service conductors and service grounding conductors may not enter, or run through, this section.

15. GROUNDING AND BONDING

GROUNDING CONNECTIONS IN SERVICE PANEL

Rule 10-204 - The service grounding conductor must enter the service box at the correct location in the service section of the box, and be correctly connected to the neutral pad as shown below.

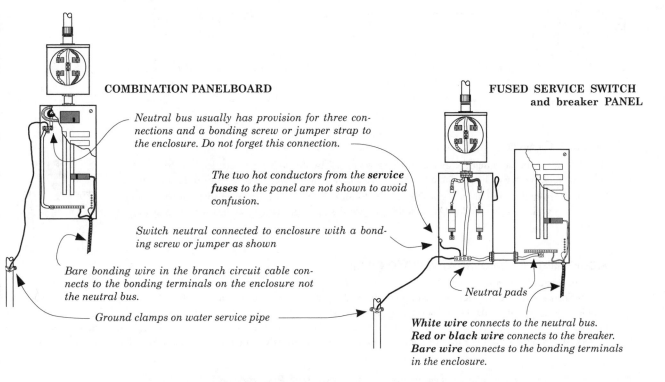

COMBINATION PANELBOARD

Neutral bus usually has provision for three connections and a bonding screw or jumper strap to the enclosure. Do not forget this connection.

*The two hot conductors from the **service fuses** to the panel are not shown to avoid confusion.*

Switch neutral connected to enclosure with a bonding screw or jumper as shown

Bare bonding wire in the branch circuit cable connects to the bonding terminals on the enclosure not the neutral bus.

Ground clamps on water service pipe

FUSED SERVICE SWITCH
and breaker PANEL

Neutral pads

White wire connects to the neutral bus.
Red or black wire connects to the breaker.
Bare wire connects to the bonding terminals in the enclosure.

SERVICE SWITCH OR COMBINATION PANEL

Rule 10-204 - The illustrations above show bonding in the service box. The illustration above left shows the bonding screw in the service section of a combination panel and the illustration on the right shows a bonding jumper in a service switch. All service equipment is provided with a brass bonding screw, or a bonding jumper, which must be installed to connect the neutral bus in the main section to the metal enclosure. Where a bonding jumper is used, for example, perhaps a bonding screw or strap wasn't provided with the electrical equipment, the bonding jumper must be sized according to Table 16A/B in the Code. This is a very important point. Without this connection a short circuit on any of the branch circuits would not trip the breaker supplying that circuit and all exposed metal parts on that branch circuit would become hot and very dangerous. Do not forget this detail in a main service but, in any sub panel, such as shown on the right where the panel is a separate box, the brass screw or jumper must not be installed.

GROUND CABLE

- Rule 10-812 - Ground cable runs from the panel to the grounding electrode.
- Cable type required = Copper and aluminum are both permitted (10-812), but copper is easier to work with and is the usual choice.
- Insulated or bare = It may be insulated or bare. If it is insulated the insulation must be green, or green with one or more yellow stripes, throughout its entire length. Rule 4-036(1) & (2) will permit a #2 and larger insulated bonding and grounding conductor to be painted green. Check with your Inspector before painting an insulated bonding or grounding conductor which is smaller than #2. (10-802)
- Size of cable = #6 It's best to use #6 cable for all services up to and including the 200 amp service because this size cable may be run exposed where not subject to severe mechanical injury.

- Protection required = The #6 copper cable may be used where it is not subject to mechanical injury. Wood moulding or plastic pipe, may be used to protect those portions which may be subject to mechanical injury.
- Cable Support = This bare cable may be stapled in place using loomex cable staples.

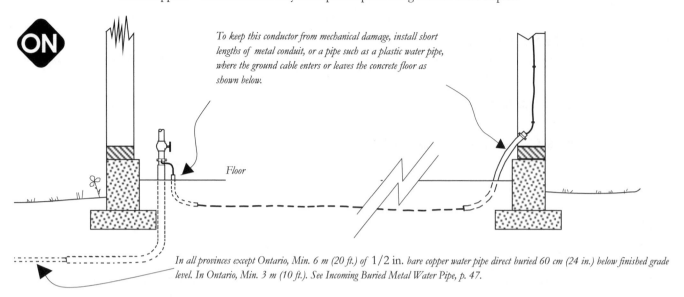

To keep this conductor from mechanical damage, install short lengths of metal conduit, or a pipe such as a plastic water pipe, where the ground cable enters or leaves the concrete floor as shown below.

Floor

In all provinces except Ontario, Min. 6 m (20 ft.) of 1/2 in. bare copper water pipe direct buried 60 cm (24 in.) below finished grade level. In Ontario, Min. 3 m (10 ft.). See Incoming Buried Metal Water Pipe, p. 47.

GROUND CLAMPS

Rule 10-908 - Make sure that the ground clamps you install are not only CSA certified but also that they are of copper, bronze or brass. The dry type connectors will not be approved in any outdoor location.

Where the ground clamps are in a consistently dry location, such as in a dry crawl space or basement or located in a wall, as shown under "Accessibility" below, the dry type clamp may be used. It's easy to get caught on this one - watch it.

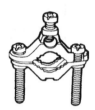

ACCESSIBILITY TO GROUND CLAMP CONNECTIONS

Rules 10-902(2) & 10-904(2) - Rule 10-902(2) says that if the electrical service is grounded to the incoming water service pipe, as described above, the connection must remain accessible wherever practicable. Rule 10-904(2) says the same thing about connections to ground rods. This means just what it says - leave it accessible - if it is practical to do so. If the rods or the water pipe are located in a wall, provide an access panel as shown. Ground rods are considered a tripping hazard therefore they must be recessed into the ground or in the concrete floor as shown or directly buried in the ground. See "Ground Clamps" above.

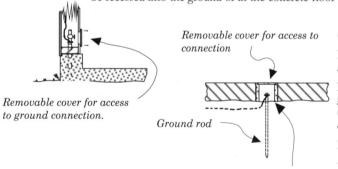

Removable cover for access to ground connection.

Removable cover for access to connection

Ground rod

Wood frame or short section of pipe set flush with surface of sidewalk or grade.

GROUNDING ELECTRODES

Rules 10-002, 10-106, 10-700 & 10-702 - New Code changes - A number of significant changes were made in the new Code but for the most part they do not affect grounding for single family houses. Most of the grounding electrodes used in the past are still acceptable under the new Code.

The object of grounding for the electrical wiring and equipment in your house is to reduce the possibility of an electric shock and fire damage. A very important reason for good grounding is to provide a good path for fault current to flow freely. High fault (short circuit) current flow is essential for a fast response from the circuit breaker which supplies the circuit. Low current flow to a short circuit means it will take longer to trip the breaker and open the circuit. Short circuits usually generate a great deal heat in a very short period of time. The concern then, is fire. Good grounding and bonding means that a good path has been provided for current to flow freely and that determines how quickly the circuit breaker will recognize the problem and open the circuit. Grounding and bonding is a very important part of your installation. Good grounding and bonding depend on a good grounding electrode.

Grounding electrodes we can use - The rules permit the following different kinds of grounding electrodes: These are not listed in any order of preference, each one is acceptable, there is no order of preference. Choose the one which best suits your need.

- Incoming Buried Metal Water Pipe - See p. 47.
- Metal Water Well Casing - See p. 47.
- Just Plain old Ground Rods - See p. 48.
- Concrete encased grounding electrode, see p. 49.
- Bare copper conductor direct buried in earth, see p. 51.

- Plate electrodes, see p. 51.

Please note - This list is not in any order of value, or effectiveness, or priority.

INTERCONNECTION REQUIRED

Rule 10-700(2) says that where both #1 & #2 type electrodes, listed above, are available we must connect them together with a copper conductor the same size as for the main service ground. For artificial grounding electrodes, types #3 & #4 listed above, we must connect them together with a #6 copper conductor. If you have more than one electrode use the wiring methods and the means of connection to each electrode as described below.

INCOMING BURIED METAL WATER PIPE

Rule 10-700(1)(c) - The Rule calls this an "in-situ" grounding electrode. An incoming copper water service pipe is acceptable provided it is buried at least 60 cm (24 in.) below finished grade and that it has an exposed bare metal area equal to that provided by, for example, two CSA certified steel ground rods, Rule 10-700(2) & (4).

Rule 10-700 says that whenever we have at least 10 ft. (3 m) of continuously conductive metal water piping system that is located underground at least 60 cm (24 in.) below finished grade level entering a single family house we must use that metal pipe as the service grounding electrode. This arrangement is acceptable without any additional artificial grounding electrodes.

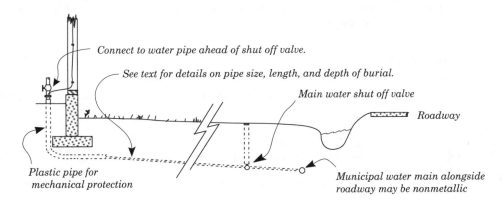

Connect to water pipe ahead of shut off valve.

See text for details on pipe size, length, and depth of burial.

Main water shut off valve

Roadway

Plastic pipe for mechanical protection

Municipal water main alongside roadway may be nonmetallic

GAS LINES

Both natural gas and propane gas piping - Rule 10-406(4) (& Bulletin 10-14-4 - in Ontario) Metal gas piping must be bonded to the service grounding electrode. Yes, most of these gas burning devices are also connected to the electrical system and are bonded to ground with a bonding conductor in their supply cable but the rule still says bond it again. This may not make a whole lot of sense but that is what the rule says.

The gas piping system is normally bonded to ground at only one point as described for water pipe bonding on p. 45 . However, short sections of isolated metal gas piping to appliances must also be bonded to ground. The service grounding conductor may continue on from the grounding electrode to pick up the cold water piping system, the gas piping system and, if in an old house with cast iron plumbing, a waste piping system. Do not cut the grounding conductor at the connectors, simply run through it and on to the next bonding clamp as shown for a furnace on p. 118

Caution - the standard ground clamp shown, above left, may damage soft copper tubing used in some gas distribution systems. Where the gas lines are of soft copper, Rule 10-614(2) permits the use of a copper strap wrapped around the pipe and clamped tightly with a bolt as shown. A #6 copper grounding conductor is then connected to this strap with a connector lug bolted (with another separate bolt) to the strap. The other end of this #6 copper conductor is then connected to the nearest grounded metal cold water pipe. This grounding conductor must be installed with the same care and attention you would use in installing a service grounding conductor.

Use 2 bolts - do not try to do this with one bolt.

This strap may be used for bonding gas tubing to the service ground.

METAL WATER WELL CASING

The Rule calls this an "in-situ grounding electrode." Rule 10-700(1)(c) permits a metal water well casing to be used as the service grounding electrode. This too must have at least 462 square in. of bare metal exposed to direct contact with earth at least 60 cm below final grade.

Although the above Rule permits a metal water well casing to be used as the service grounding electrode, however, the Ontario Ministry of Environment will not permit it, Bulletin 10-16-1. Therefore we must find a grounding electrode that will satisfy the rules of both the Ontario Electrical Code and the Ontario Ministry of the Environment.

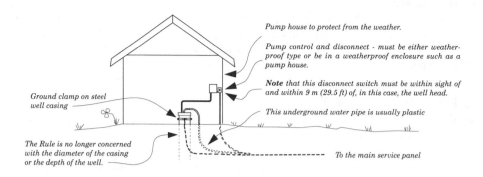

Pump house to protect from the weather.

Pump control and disconnect - must be either weather-proof type or be in a weatherproof enclosure such as a pump house.

Note *that this disconnect switch must be within sight of and within 9 m (29.5 ft) of, in this case, the well head.*

Ground clamp on steel well casing

This underground water pipe is usually plastic

The Rule is no longer concerned with the diameter of the casing or the depth of the well.

To the main service panel

Metal well casings usually provide much more bare metal contact with earth than is required. Note that there is no minimum casing length or diameter required now. All that is required is the minimum bare metal contact with earth and that contact area be at least 60 cm below final grade level.

GROUND RODS

The Rule calls this a "manufactured grounding electrode." Where there is no continuously conductive metal water piping system available we may use ground rods. The following picky points must be observed when installing ground rods.

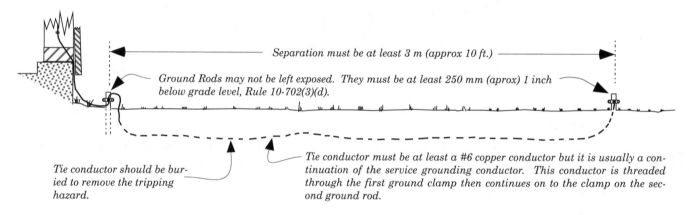

Separation must be at least 3 m (approx 10 ft.)

Ground Rods may not be left exposed. They must be at least 250 mm (aprox) 1 inch below grade level, Rule 10-702(3)(d).

Tie conductor should be buried to remove the tripping hazard.

Tie conductor must be at least a #6 copper conductor but it is usually a continuation of the service grounding conductor. This conductor is threaded through the first ground clamp then continues on to the clamp on the second ground rod.

GROUND ROD SURFACE AREA

The standard ground rod has a diameter of 5/8 in. and is 3 m (118 in.) long and let's remember Rule 10-700(2)(a) requires two such rods for an acceptable grounding electrode.

The total surface area of a standard ground rod is 231 square in. Two rods would have a surface area of 462 square in. and that is the minimum bare surface area required for copper pipe to be accepted as a grounding electrode.

• 1/2 in. copper pipe - 609 cm (20 ft) long has a surface area of 462 square in. and is acceptable.
• 3/4 in. copper pipe - 499 cm (17 ft.) long has a surface area of 462 square in. and is acceptable.

If we want to use the water service pipe as our grounding electrode these are the minimum lengths of bare copper pipe which must be buried at least 60 cm (24 in.) below grade level. If this bare section is not available with the water service pipe we must choose one of the other grounding electrode options listed above.

CSA CERTIFICATION

Ground rods are now required to be properly certified for use by CSA or by one of the other approved certification agencies. See the partial list of these agencies, p. 2. This also applies to plate electrodes. Home made rods or plates are no longer acceptable, the Inspector will be looking for that certification label.

MINIMUM SIZE

The Code no longer specifies either length, diameter, or kind of material required for a ground rod. It is expected they will be the standard size which we are all familiar with, 3 m (10 ft.) by 1.58 cm (5/8 in.). The same is true for plated electrodes. If the rod or the plate is properly labeled, it's okay to use it.

Number of Rods Required

Rule 10-700(2)(a) required at least two rods in every case, but we may actually need more in some special cases in order to provide an adequate ground return path for fault currents. Your Inspector will advise you if more than two rods are required.

Depth

Ground rods must be driven into the ground their full length plus one in. as shown above.

Spacing

Ground rods must be spaced at least 3 m (10 ft.) apart.

Tie Conductor

The rods must be connected together with a #6 or larger, copper bonding conductor. Normally, this tie conductor is a continuation of the service grounding conductor. Simply thread the conductor through the first rod connector and continue on to the second rod connector.

Ground Clamps

These connectors must be of copper or bronze, if they are placed in any wet or damp location. Dry type connectors may be used only indoors in dry locations.

Tripping Hazard

In the illustration above the ground clamp connections to the rods are buried below grade level. This is required by Rule 10-700(2)(a)(2) to remove the tripping hazard. The same is required for the tie cable between rods to remove the tripping hazard.

Rock Bottom

Can't drive that thing into the ground? There will be locations where the ground cover is less than 3 m (10 ft.) and the ground rods can not be driven in the usual way. The rules allow the following labour intensive solutions. Choose one that best fits your situation.

Rule 10-700(5) says "Where a local condition such as rock or permafrost prevents a rod or a plate grounding electrode from being installed at the required burial depth, a lesser acceptable depth shall be permitted". The Rule does not say so, but any depth less than 600 mm (24 inches) could mean that you are in deep trouble, (no pun intended)

The solution is to find a location where there is at least 600 mm (24 in.) of ground cover. This will permit the direct burial of a ground rod in the horizontal position as shown in Method A. In Method B there is more depth so that the rod could be driven into the ground at an angle. Method C shows a ground rod driven into the ground as far as possible then bent over. Bending a 5/8 inch diameter ground rod without first heating it is difficult in a shop. Bending it when held in only a few feet of earth would be even more difficult and would greatly compromise it's effectiveness because it would no longer be held in intimate contact with earth. This is an acceptable method but it is a poor method.

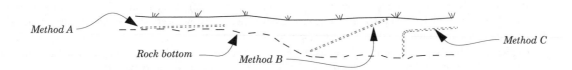

The grounding conductor from the service panel to the rod should be buried to remove the tripping hazard. Where this ground cable must run across bare rock Rule 12-012(7) requires a trench 150 mm (6 in.) deep for the cable, then grouted with concrete to the level of the rock surface.

Concrete Encased Grounding Electrodes

The Code calls this a "field assembled grounding electrode." It consists of a single copper conductor, #4 or larger, encased in the concrete footings of the building. The copper conductor is thus in intimate contact with the concrete and thence with the earth under the footings. Properly installed these electrodes are very effective. Herein lies the key - to ensure an acceptable installation your Inspector may want to see it before they will permit it to be covered. Check with your Inspector.

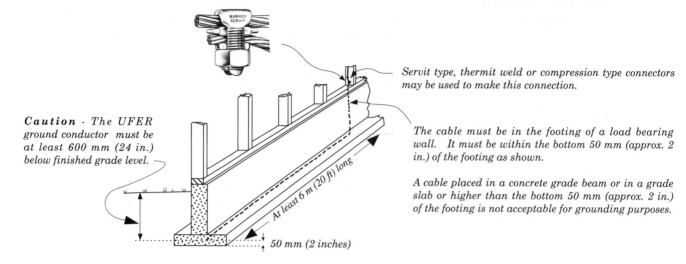

Servit type, thermit weld or compression type connectors may be used to make this connection.

Caution - The UFER ground conductor must be at least 600 mm (24 in.) below finished grade level.

At least 6 m (20 ft) long

50 mm (2 inches)

The cable must be in the footing of a load bearing wall. It must be within the bottom 50 mm (approx. 2 in.) of the footing as shown.

A cable placed in a concrete grade beam or in a grade slab or higher than the bottom 50 mm (approx. 2 in.) of the footing is not acceptable for grounding purposes.

Conductor Required

Type
• Must be copper. .
• Must be bare.

Length - Must be at least 6 m (approx. 20 ft.) but may be much longer. This is the minimum horizontal length of conductor along the base of the footing. In addition to the 6 m length of conductor run horizontally along the bottom 5 cm (approx. 2 in.) of the footing, you will require enough length to run up to the top of the foundation wall. Allow an additional 6 in. (approx. 15 cm) or so for connection to the service grounding conductor. See illustration above.

Size - Note the following grounding conductor sizes for both the portion which is concrete encased and the part above the concrete, the home run to the panel. This cable may be spliced above the concrete foundation level as shown in the illustration above.

Ampacity of Service conductors	Concrete encased portion Code Table 43
0 to 100 amps	#4 copper
101 to 125 amps	#4 copper
126 to 165 amps	#4 copper
166 to 200 amps	#3 copper
201 to 260 amps	#2 copper

Position - This grounding conductor must be in the bottom 2 in. (5 cm) of the footing. It is not correct to place it in the floor slab or under it even if it is encased in concrete. The effectiveness of this electrode depends, in part, on the weight of the building to maintain intimate contact with earth under the footing.

Connection - It is best to use one continuous conductor and run it all the way back to the service panel but there is nothing wrong with a splice as shown in the illustration. Use a split bolt type connector to make the splice.

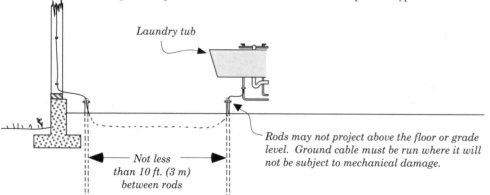

Laundry tub

Not less than 10 ft. (3 m) between rods

Rods may not project above the floor or grade level. Ground cable must be run where it will not be subject to mechanical damage.

Plastic Water Service Pipe

Rule 10-406(2) & 10-702(3) - Whenever the water service is with plastic pipe we cannot use it as our grounding electrode. We must in those cases install an artificial grounding electrode. Don't forget the bonding requirement. Even if the water service is plastic pipe, the piping system inside the building may be metallic. If it is, it must be bonded to the ground rods, Rule 10-406(2). This connection may be made at any convenient point on the cold water pipe where it will remain accessible.

Bare Copper Conductor Direct Buried in Earth

Rule 10-700(3)(b) calls this a "field-assembled grounding electrode." This is a new option, it has never been part of the Rules before.

Conductor size required - These sizes are exactly the same as for a concrete encased grounding electrode described above.

This may not be a popular option because it involves digging a 6 m (20 ft.) trench 60 cm (24 in.) deep. This cable could be located in a trench used for other underground runs. Since the grounding electrode is such an important part of the installation it will likely need to be inspected before it is covered in. Confirm this with your inspector.

Plate Electrodes

Rule 10-700(2)(b) - The Code calls this a "manufactured grounding electrode." This means that plate electrode must be manufactured to a rigid CSA Standard and be marked to show they are certified for use as a grounding electrode. The Rule does not specify any physical dimensions because we are not allowed to make the plate ourselves. Check for that certification label. The total surface area of the plate exposed to direct contact with earth or as shown, completely buried in concrete, must be at least 0.2 m^2. It must be not less than 6 mm thick if of iron or steel; or 1.5 mm thick if of non-ferrous metal; and the plate must be buried at least 60 cm below finished grade level or it may be located in the concrete footings of a foundation as illustrated above. In either case the Rule requires the plate to be at least 60 cm below final grade level. Where this plate is buried in concrete it must be completely encased in the bottom 2 in. layer of the footing. CSA certified grounding plates are available now.

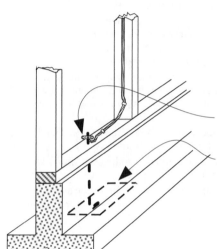

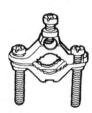

Use ground cable connector.

This is a manufactured and certified steel plate electrode. This plate can be:

    a -   *Direct buried **in the ground** but must be at least 600 mm (24 inches) below grade level, or.*

    b -   ***Encased in concrete**, as shown, where it must be at least 600 mm below grade level and it must be encased within the bottom 50 mm (approx. 2 inches) of a concrete foundation footing which is in direct contact with earth.*

Bonding Required

Rule 10-406(2)(3)(4) Don't forget the bonding requirement. Even if the water service is plastic pipe, the piping system inside the building may be metallic. All metal piping for such as water, gas, or waste systems in the building, which are not used as a grounding electrode, must be bonded to ground with a #6 copper conductor. This conductor must be installed with the same care and attention to detail required when installing the service grounding conductor. This conductor connects to each of the metal piping systems with a regular ground clamp or a grounding strap as shown in the illustration, then it connects to the service grounding conductor at any convenient point which will remain accessible for inspection when future load changes are made.

Bonding Straps - Rule 10-614(2) - This Rule permits a home made bonding strap. The minimum dimensions must be:

• 19 mm x 1.4 mm for a strap made of steel.
• 19 mm x 1.2 mm for a strap made of copper.

Converting these dimensions into familiar terms:

• 19 mm = 0.75 in. the strap must be at least 3/4 in. wide, either steel or copper.
• 1.4 mm = 0.055 in. thick if the strap material is steel, it must be at least 16 gauge.
• 1.2 mm = 0.047 in. thick if this strap material is copper or aluminum it must be at least #18 gauge.

Use 2 bolts - do not try to do this with one bolt.

This strap may be used for bonding gas tubing to the service ground.

16.   BRANCH CIRCUIT WIRING

GENERAL INFORMATION

First, some definitions used in the Electrical Code and in this book.

- Wire - This refers to a single conductor. For example, it refers to one of the two, or three, single conductors in a 2-wire, or 3-wire cable. When a three wire cable enters an outlet box there are then three wires in that box, not four, the bare wire does not count.

- Cable - The word cable usually means two or more conductors in a manufactured cable assembly. A large, single conductor is also referred to as a cable but in that case it is always identified as a single conductor cable.

- Colour Coding - Rule 4-036 - The colour of individual insulated conductors, not cables, is very important. Cables can have any colour except White, Grey, or Green, but the individual conductors in the cable must follow the Code as shown below.

  The Neutral conductor is white, grey, or in flexible cable it is often identified with a raised longitudinal ridge. Conductors with white or grey coloured insulation, those with a raised longitudinal ridge and those with a white tracer ribbon must not be used for any purpose other than a neutral conductor except as noted on p. 68 when used for switch legs.

  The Bonding Conductor, may be bare but if it is insulated, must be green. A conductor with green insulation must not be used for any purpose other than for bonding. I suppose it could be used for tieing tea bags and such things.

  The ungrounded conductors - The hot conductors - In single family residential wiring this refers to all the other conductors in a cable, outlet box, or control panel. These can be any colour other than white, grey or green.

- Receptacle - Code Section 0 - These are generally identified as a single receptacle which has one female contact device, or a duplex receptacle which has two female contact devices. When counting the number of outlets on a circuit the single receptacle and the duplex receptacle each count as one outlet on the branch circuit. Your local bank manager does not agree with this, but please accept this generosity from the gentle folk who planned this Code.

- Outlets per circuit- Rule 8-304 permits 12 outlets on a circuit. It is better to supply only about eight or nine outlets on a circuit and these should be a mixture of light outlets and plug outlets. When counting outlets on a circuit count only those outlets which supply a load, such as a fixed lighting fixture, or a plug outlet which can supply a load. For example do not count a switch box or a junction box.

- Lights & Plugs - It is better to have both general lighting and general plug outlets on the same circuit rather than having these on separate circuits. When a circuit fails a smaller portion of the house is plunged into darkness.

- Split receptacle - This refers to a duplex receptacle equipped with a break-away tab so that it can be used with 3-wire cable where each of the two female contact devices are supplied with a separate circuit. See the illustration on p. 82.

- Tamper resistant plugs - This is the new kid on the block. All 15 amp and 20 amp plug outlets both inside and outside of a single family house must be tamper resistant type except those plug outlets dedicated for microwaves, refrigerators, freezers, plugs on the kitchen counter and those receptacles located in an attic or crawl space. All others must be Tamper resistant type receptacles.

- Lighting circuit - Maximum fuse or breaker rating for a lighting circuit in a residence is 15 amp.

- Plug outlet circuit - breaker and supply cable rating must be equal to the rating of receptacle - either 15 amp or 20 amp. Most outlets must be supplied with a special circuit breaker called an AFCI (arc-fault circuit interrupter), see p. 77. The exceptions are: kitchen counter, island and peninsula receptacles, the refrigerator receptacle and the receptacle for a sump pump, provided it is a single receptacle and is labelled as being for the sump pump receptacle. Note that the receptacles themselves on an AFCI circuit can be GFCI type.

- Kitchen Counter Plugs - These require special care. Any counter plug outlets within 1.5 m (59 in.) of the kitchen sink must be GFCI protected. We can use the familiar 15 amp 3-wire split receptacle for those kitchen counter plug outlets where GFCI protection is not required. These circuits can also be used for kitchen counter plugs located in the GFCI zone provided we use a 2-pole circuit breaker in the panel instead of two single pole breakers with tie bar. Because 15 amp 2-pole GFCI type circuit breakers are expensive, you may want to use the 20 amp single pole GFCI type breaker with 20 amp cable and 20 amp T-slot type receptacle instead.

- Range and Dryer Plug Outlets - Rule 26-744, see also "Heavy Appliances," p. 114.

NUMBER OF CIRCUITS

Rules 8-108, 8-304 & 26-722 - Each circuit breaker or fuse may supply only one circuit. It is not correct to connect two or more wires to a circuit breaker or fuse, Rules 6-212 & 12-3032. A sufficient number of breaker or fuse spaces should be provided in the service panel to comply with this requirement. See p. 41 under "Minimum Circuits Required."

FUSES

Rule 14-204 - Where fuse type panels are used it is difficult, if not impossible, to control the size of branch circuit fuses used. It is too easy to replace a blown fuse with one of a higher rating. To avoid this, the Code requires that all plug fuse holders must be equipped with a rejection feature that will make it impossible to replace a fuse with one of a higher rating. The illustration on the left shows a fuse and a rejection washer. The washer is inserted into the fuse socket. Each washer has a different size opening which will prevent a fuse of higher rating from making contact. Do not remove these washers.

Size of Cable

Use only #14 wire unless your runs are unusually long, say more than 40 m (approx. 130 ft.) long. Exception, Kitchen plugs rated for 20 amps must be wired with #12 copper wire. The #12 wire is stiff; it can cause excessive strain on the switch and receptacle terminals and may require larger outlet boxes. See table on p. 82.

3-Wire Cable

You are running two circuits in each cable. Rule 14-010(b) now says that we do not need tie bars or two pole circuit breakers for 3-wire cables except where these cables supply 240 volt loads, such as a range or dryer etc. or 120 volt split receptacles. This means that two single pole circuit breakers, without a tie-bar are acceptable for a 3-wire cable which supplies only lighting or plug outlets each of which is connected to the neutral and one hot conductor.

Exception - 3-wire cables used to supply any split receptacles, such as are used on the kitchen counter, must be protected with either a two pole circuit breaker or with two single pole breakers which have their operating handles connected together with a tie-bar. This is to ensure that both circuits are de-energized for safety for anyone working on these special outlets. If you are using a fuse panel you will need a special fuse block and fuse pull for these special circuits. See p. 37 for details on 2-pole breakers and tie-bars.

For heavy appliance wiring, look under specific type.

Cables Bundled Together

Rule 4-004(12) - This rule says that where cables are "in contact with each other for distances exceeding 60 cm (24 in.) the ampacity of the conductors must be corrected by applying the factors of Table 5C."

Table 5C in the Code

1 to 3 conductors may carry .... 100% load
4 to 6 conductors may carry .... 80% load
7 to 24 conductors may carry .... 70% load
25 to 42 conductors may carry .... 60% load
43 or more conductors may carry .... 50% load

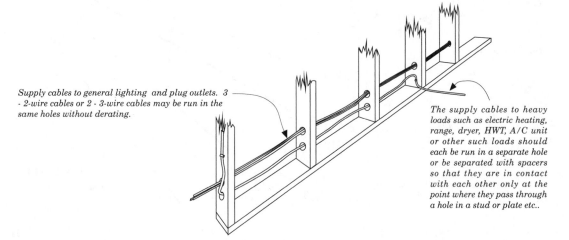

Supply cables to general lighting and plug outlets. 3 - 2-wire cables or 2 - 3-wire cables may be run in the same holes without derating.

The supply cables to heavy loads such as electric heating, range, dryer, HWT, A/C unit or other such loads should each be run in a separate hole or be separated with spacers so that they are in contact with each other only at the point where they pass through a hole in a stud or plate etc..

A #14 loomex cable is normally permitted to carry 15 amperes but when it is run bundled together with other cables it may have to be derated to 80% or even 70% depending on the number of conductors, (not cables) in the bunch.

Logic - Before the days of super insulated and super sealed houses cables could be bundled together without concern. There was usually sufficient air movement in the walls and the ceiling to prevent dangerous temperature rise in the bundle due to mutual heating. Today we build fully sealed sardine cans which are highly insulated against heat loss. For example, the BTUs your furnace produced in the fall cannot escape until you open the doors in the spring. Under these conditions cables bundled together could overheat unless they are derated so that each conductor carries a smaller load.

The rule is aimed at cables which supply electric heat, range, dryer, water heater and any other heavy loads such as an A/C unit. The concern is not with cables supplying general residential lighting and ordinary plug outlets although the rule could be applied to these cables as well.

CHART OF BRANCH CIRCUITS

Alert! Most branch circuits now require AFCI. See "Plug outlet circuit," p. 52, and "Plug Outlets," starting on p. 74.

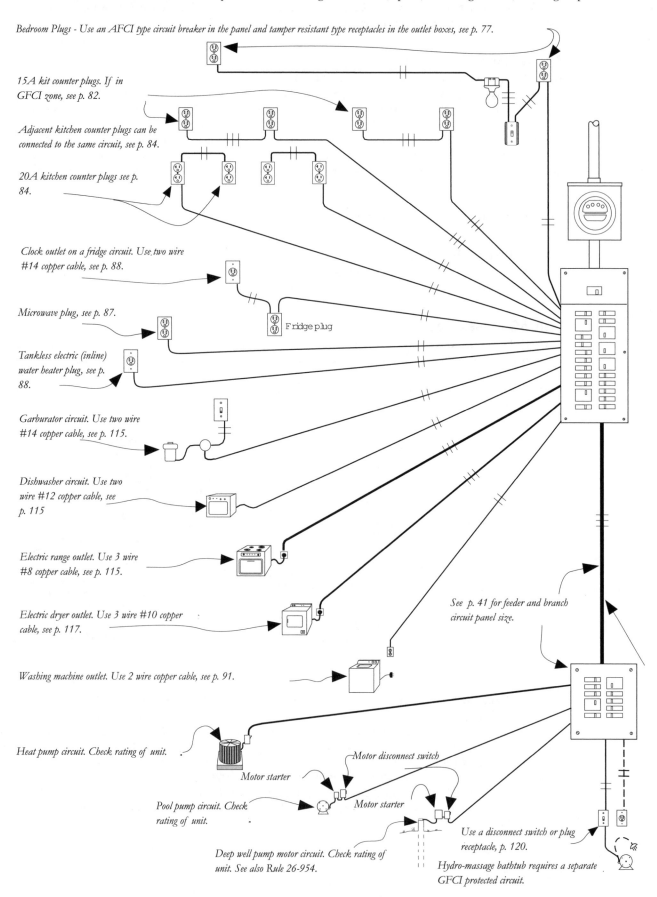

Bedroom Plugs - Use an AFCI type circuit breaker in the panel and tamper resistant type receptacles in the outlet boxes, see p. 77.

15A kit counter plugs. If in GFCI zone, see p. 82.

Adjacent kitchen counter plugs can be connected to the same circuit, see p. 84.

20A kitchen counter plugs see p. 84.

Clock outlet on a fridge circuit. Use two wire #14 copper cable, see p. 88.

Microwave plug, see p. 87.

Fridge plug

Tankless electric (inline) water heater plug, see p. 88.

Garburator circuit. Use two wire #14 copper cable, see p. 115.

Dishwasher circuit. Use two wire #12 copper cable, see p. 115

Electric range outlet. Use 3 wire #8 copper cable, see p. 115.

Electric dryer outlet. Use 3 wire #10 copper cable, see p. 117.

See p. 41 for feeder and branch circuit panel size.

Washing machine outlet. Use 2 wire copper cable, see p. 91.

Heat pump circuit. Check rating of unit.

Motor disconnect switch

Motor starter

Pool pump circuit. Check rating of unit.

Motor starter

Use a disconnect switch or plug receptacle, p. 120.

Deep well pump motor circuit. Check rating of unit. See also Rule 26-954.

Hydro-massage bathtub requires a separate GFCI protected circuit.

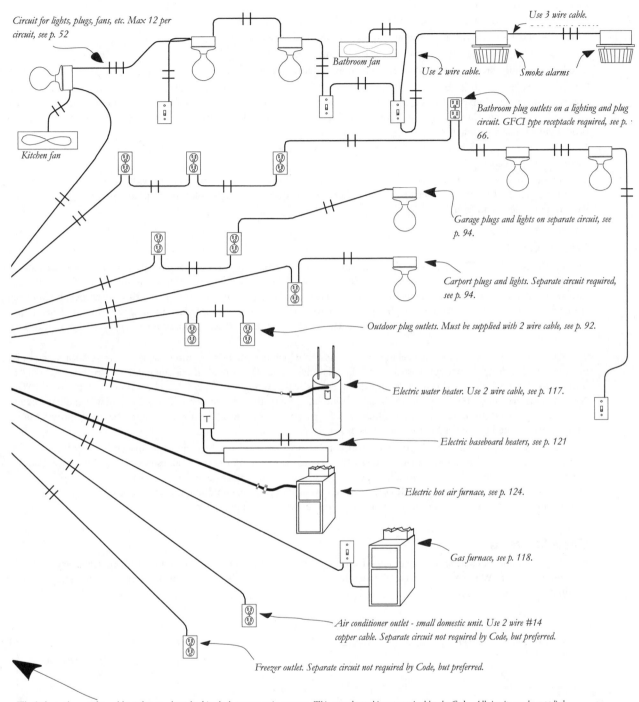

Circuit for lights, plugs, fans, etc. Max 12 per circuit, see p. 52

Kitchen fan

Bathroom fan

Use 2 wire cable.

Use 3 wire cable.

Smoke alarms

Bathroom plug outlets on a lighting and plug circuit. GFCI type receptacle required, see p. 66.

Garage plugs and lights on separate circuit, see p. 94.

Carport plugs and lights. Separate circuit required, see p. 94.

Outdoor plug outlets. Must be supplied with 2 wire cable, see p. 92.

Electric water heater. Use 2 wire cable, see p. 117.

Electric baseboard heaters, see p. 121

Electric hot air furnace, see p. 124.

Gas furnace, see p. 118.

Air conditioner outlet - small domestic unit. Use 2 wire #14 copper cable. Separate circuit not required by Code, but preferred.

Freezer outlet. Separate circuit not required by Code, but preferred.

The feeder to the second panel located near a large load in the basement or in a garage. This second panel is not required by the Code. All circuits can be supplied from one main panel, see p. 41.

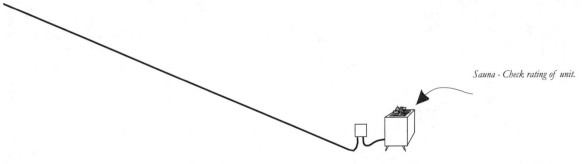

Sauna - Check rating of unit.

SOLUTION

Cables supplying general lighting and plug outlets have always been derated to 80% by Rule 12-4,000 and Table 5C permits 6 conductors bundled together with this 20% reduction in current carrying capacity. Therefore, cables supplying these outlets may be bundled as follows:

• Three 2-conductor cables may be run in contact without further derate, or

• Two 3-conductor cables may be run in contact without further derate

All other cables should be run in separate holes to ensure separation

The bare bonding wire in the cables is not counted in this application.

The derating applies only where cables are in continuous contact with each other for distances greater than 24 in. Where the contact is less than 24 in. the derating does not apply. This means you could have many cables running through a hole in a stud without derating provided that the cables then fan out in different directions so that they are not in physical contact for more than 24 in. at any one point.

If you fail to observe this rule your inspector may need to ask you to restring the cables in different holes, or worse, replace them with larger cables.

The simplest solution is to force a bit of insulation or a wood chip between the cables to provide the required separation. Even though the cables are in contact with each other where they pass through holes the continuous contact in each case would then be less than 24 in.

TYPE OF CABLE

Rules 2-126 & 30-408 - FT1 Cable marking - All loomex cables used in any wiring in any building must have at least an FT1 marking. This mark shows the cable insulation has been properly tested for burn-out. That is, in the event of a fire the cable insulation will not be the means by which the fire can travel from one part of the house to another. Your local building supply store will not likely have unmarked cable but you may have some left over from that last bit of wiring you did. Check that old cable. If the FT1 marking is not shown the cable is usable only for non-electrical applications such as fencing wire, or as a clothes line, etc.

90° Insulation on loomex Cable - All ceiling outlet boxes on which it is intended to mount a light fixture, (not junction boxes) must be wired with 90°C conductors such as NMD90 loomex cable. This sounds threatening, but the truth is it would be difficult to find any loomex cable intended for dry locations with a rating less than 90°C. There is one exception. NMW and NMWU cable has only a 60°C rating. NMW cable is intended for wet locations such as barns etc. NMWU cable is intended for use underground. In these damp or wet locations the rule permits these 60°C cables to enter a ceiling light outlet box which is also located in the damp or wet location.

Cold Regions - Rule 12-102, - Insulated conductors such as NMD90 cable can be seriously damaged if it is flexed at temperatures lower than -10 degrees C (14 degrees F). Do not install cables in cold temperatures unless the cable is specifically approved for use at that temperature. Failure to observe this requirement may result in a seriously compromised installation. It's no fun doing this work in cold weather anyhow. This work should be done in warm weather when it can be fully enjoyed.

CABLE STRAPPING

Rule 12-510 - Loomex cable should be strapped within 30 cm (12 in.) of the outlet box and approximately every 1.5 m (59 in.) throughout the run. Where cables are run through holes in studs or joists they are adequately supported and strapping is not required.

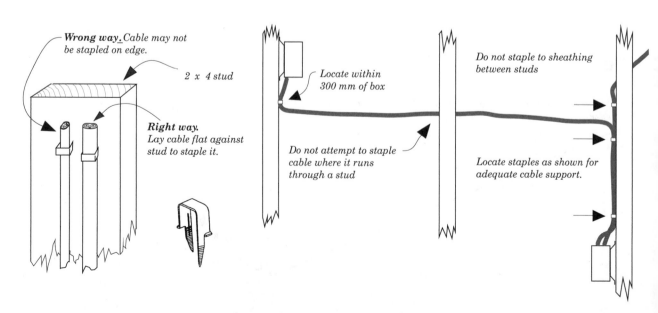

Wrong way. Cable may not be stapled on edge.

2 x 4 stud

Right way. Lay cable flat against stud to staple it.

Locate within 300 mm of box

Do not attempt to staple cable where it runs through a stud

Do not staple to sheathing between studs

Locate staples as shown for adequate cable support.

CAUTION

(1). Do not staple 2 wire cable on edge - they must lie flat. See Rule 12-506(5). This applies to two wire cables only.

(2). Do not overdrive the staples. Drive staples only until they contact the cable sheath - don't squish the cable. Because the cables are scantily dressed (insulated) the installers must be more careful when handling and strapping it.

(3). Rule 2-108 - Be sure to use the correct size staple or strap for each size cable. It is not correct to use a 2-wire cable strap on a 3-wire cable unless the staple or strap is specifically approved for both sizes, nor is it correct to put two cables under a single strap or staple. You may get away with two cables under a strap if they are very carefully installed.

(4). Where cables are run along studs or joists they should be kept at least 1-1/4 in. from the nailing edge. Between staples the cable is free to move aside should a dry-wallers nail miss the stud but at the point of the staple the cable is held captive. If the cable has been stapled too close to the edge of a stud or joist it really needs protection. There is no code rule which specifically requires this protection except that Rule 2-108 says poor workmanship will not be accepted by the inspection department. It is best to run your cables along the middle, or as near the middle, of the stud or joist wherever possible and provide additional protection where it is not possible to run it there.

CABLE PROTECTION

Rule 12-516 - Where the cable is run through holes in studs, plates or joists, these holes must be at least 32 mm (1-1/4 in.) from the edge of the wood member. For protection, a plate may be used as shown in the illustration below, or a cylindrical bushing may be used.

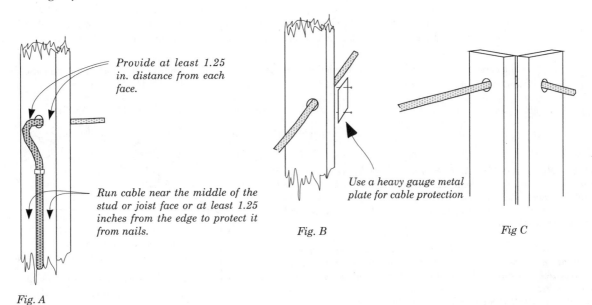

Provide at least 1.25 in. distance from each face.

Run cable near the middle of the stud or joist face or at least 1.25 inches from the edge to protect it from nails.

Use a heavy gauge metal plate for cable protection

Fig. A

Fig. B

Fig C

In the case of small dimension members such as may be used in partitions, the cable hole should be located so that there is 32 mm (1-1/4 in.) clearance on one side. Fig. B. To protect the other side use a minimum #16 MSG steel plate (the side of a metal sectional outlet box does this job very well). This must be done in every case where the 32 mm (1-1/4 in.) clearance cannot otherwise be obtained. In corners of rooms, as in Fig. C. the holes may need to be drilled at an angle providing less than the minimum distance - here too, use heavy metal plates to protect the cable from dry-wallers nails.

Holes may contain more than one cable but must be large enough to prevent damage to the cable sheath during installation.

Bundled cables could be a serious problem. Where a number of cables are run in contact with each other you may be in conflict with Rule 4-004(11) depending somewhat on the type of loads served by those cables. See "Cables Bundled Together," p. 53 for details.

KITCHEN CABINETS

Kitchen cabinets, or other similar cabinets, are often supported with long screws or nails that penetrate more than 1.25 in. into the studs. For this reason all cables not directly required in this wall space, (where cabinets may be mounted) should be kept out of this wall area or be protected with #16 MSG metal plates. The rules do not specifically require this but it may keep you out of a lot of trouble later.

Obviously some cables will need to be run in this wall space to supply the counter plugs, dishwasher, garburator, hood fan etc. but all other cables should be kept out of this wall space. Where possible, cables which must run in this wall space should be run in a vertical direction, not horizontally. Where it must be run horizontally through a stud, use heavy (#16 gauge) steel plates or the equivalent, or keep the cable at least 2 in from the nailing edge of the studs. Remember, this is not a Code rule, but it may save trouble later.

ATTIC SPACES

Rule 12-514(a) says cables may be run on the upper faces of joists or lower faces of rafters, as shown below, provided the head clearance, joist to rafter, is 1 m (39.4 in.) or less. Where the head room is greater than 1 m the cables must be run through holes or on running boards as shown below. Do not drill any holes in a manufactured structural assembly.

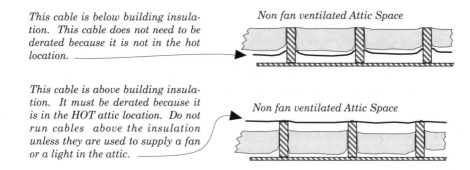

This cable is below building insulation. This cable does not need to be derated because it is not in the hot location. ———

Non fan ventilated Attic Space

This cable is above building insulation. It must be derated because it is in the HOT attic location. Do not run cables above the insulation unless they are used to supply a fan or a light in the attic. ———

Non fan ventilated Attic Space

Caution - Rule 4-004(8) Cables placed above the building insulation are in a possible high temperature location and may need to be derated to 80% of their normal ampacity. If the attic space temperature is controlled by a thermostatically controlled fan the cables may be above the insulation without derating. In a case where your insulation is blown in and your cable is run on the upper edge of the joists, the cable could be anywhere -above the insulation, in it, or under it. Under such circumstances an inspector may derate your cable. The simple solution is to supply only a small number of outlets with cables that could be subject to derating. In this way the calculated load is below the maximum permitted load for those conductors exposed to high temperature locations.

If the cable is run below the building insulation they are out of the high ambient temperature location and therefore need not be derated. Check with your local Inspector.

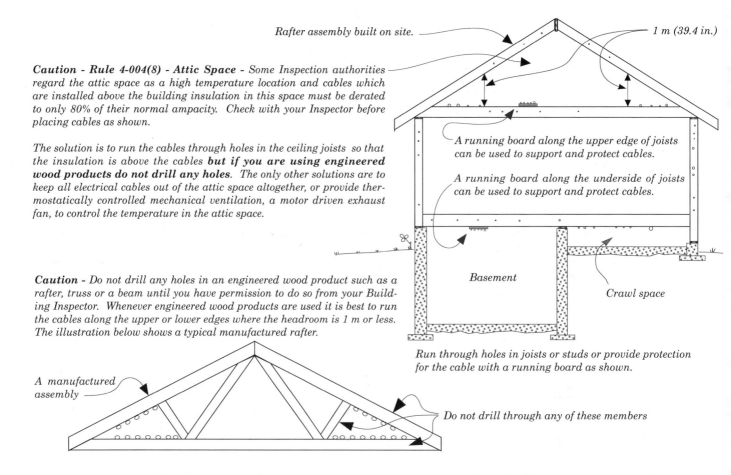

Rafter assembly built on site. ———

1 m (39.4 in.)

Caution - Rule 4-004(8) - Attic Space - *Some Inspection authorities regard the attic space as a high temperature location and cables which are installed above the building insulation in this space must be derated to only 80% of their normal ampacity. Check with your Inspector before placing cables as shown.*

*The solution is to run the cables through holes in the ceiling joists so that the insulation is above the cables **but if you are using engineered wood products do not drill any holes**. The only other solutions are to keep all electrical cables out of the attic space altogether, or provide thermostatically controlled mechanical ventilation, a motor driven exhaust fan, to control the temperature in the attic space.*

A running board along the upper edge of joists can be used to support and protect cables.

A running board along the underside of joists can be used to support and protect cables.

Basement

Crawl space

Caution - *Do not drill any holes in an engineered wood product such as a rafter, truss or a beam until you have permission to do so from your Building Inspector. Whenever engineered wood products are used it is best to run the cables along the upper or lower edges where the headroom is 1 m or less. The illustration below shows a typical manufactured rafter.*

Run through holes in joists or studs or provide protection for the cable with a running board as shown.

A manufactured assembly ———

Do not drill through any of these members

BASEMENTS AND CRAWL SPACES

Rule 12-514(b) says cables may be run on the lower face of basement or crawl space joists provided the cables are suitably protected. A running board nailed to the underside of the joists may be an acceptable protection - check with your Inspector.

It is best to protect the cables by running them through holes drilled in the joists. This allows the basement to be finished without having to re-run cables to get them out of the way.

EXPOSED CABLE

Rule 12-518 - Where loomex cable is run on the surface of the wall and within 1.5 m (59 in.) from the floor, as is often the case in buildings of solid wall construction, the cable must be protected from mechanical damage with wood or similar moulding.

HOT AIR DUCTS OR HOT WATER PIPES

Rule 12-506(4) requires loomex cable to be kept at least 25 mm (1 in.) away from all hot air heating ducts and hot water piping. A chunk of building insulation may be placed between the cable and the duct or pipe, as shown.

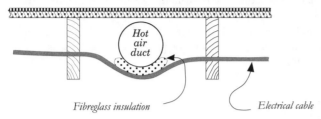

Fibreglass insulation   *Electrical cable*

COLD AIR DUCT

Rule 12-010(5) - The illustration below shows a cold air duct formed by closing in the space between joists with sheet metal. House wiring may be in this space. Outlet boxes should not be in this space unless it is necessary. In that case the outlet box must project through the sheet metal.

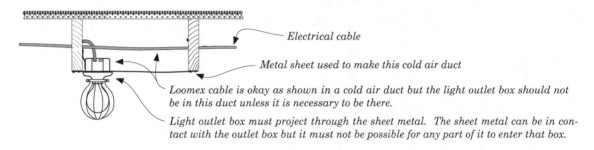

Electrical cable

Metal sheet used to make this cold air duct

Loomex cable is okay as shown in a cold air duct but the light outlet box should not be in this duct unless it is necessary to be there.

Light outlet box must project through the sheet metal. The sheet metal can be in contact with the outlet box but it must not be possible for any part of it to enter that box.

17A.   OUTLET BOXES: LIGHTS, SMOKE ALARMS, CARBON MONOXIDE ALARMS & FANS

VAPOUR BARRIER FOR LIGHT OUTLET BOXES

The Building Code requires boxes to be surrounded with a moisture resistant barrier if they are located in a wall or ceiling which is required to have a vapour barrier, see p. 65 for details.

CABLE CONNECTORS

Switch, light, plug and junction boxes normally have four separate cable entry holes for cables and each hole is equipped with a cable connector. Each of these cable entry holes is certified for one only cable entry, not two cables. Each hole may contain one 2-wire cable or one 3-wire cable. Plan, and arrange your circuit runs, so that there is no need to enter a connector with two cables.

LIGHT OUTLET BOXES

Rule 30-302(6) says rigid PVC boxes shall not be used to support lighting fixtures unless the boxes are marked to show they are suitable for that purpose. It is not clear if this refers to all plastic type light outlet boxes or only to those boxes identified and marked as PVC boxes. The problem seems to be the weakness of the material when it is subjected to the operating temperature of certain lighting fixtures. Pendant fans weighing more than 16 kg with all accessories attached must be supported by means other than the outlet box.

OUTLETS PER CIRCUIT

Rule 12-4,000 - A maximum of 12 outlets may be connected to a circuit. This may consist of 12 light outlets or 12 plug outlets (not any special outlets such as kitchen counter plugs, appliance plugs, etc. see p. 85) or any combination of light and plug outlets mixed, as long as their total number does not exceed 12 outlets. For this calculation determination, do not count junction boxes used only as junction boxes, and do not count switch outlet boxes either, and do not count appliance plugs, see p. 82.

A duplex receptacle counts as one outlet

It is better to have the load consist of a mixture of lights and plugs. This gives better load diversity on the circuit and less chance of a complete blackout in case of circuit failure.

To avoid confusion and costly duplication proceed as follows:

(1). Make a floor plan of your house. If more than one floor, draw a separate plan for each floor.

(2). Show each outlet using the symbols given below.

(3). Determine the best location for service equipment.

(4). Draw a line showing the course each cable run will take. Start with one circuit and complete it before going on to the next. Identify each circuit and each outlet for quick, easy location. For example, the outlets on the first circuit would be A1, A2, A3 etc. The outlets on the next circuit would be B1, B2, B3, etc.

It is very important that you identify the kind of circuit breaker, AFCI or GFCI or standard breaker in the panel for each circuit. Also identify the kind of receptacle, standard, GFCI, or TR, you plan to use in each outlet box. Make sure that you will have the correct protective device for each outlet.

Symbols typically used are:

 Light outlet on ceiling

 Wall light outlet such as over a bathroom vanity

$\$$ Wall switch

$\$_3$ Three way switch

$\$_4$ Four way switch

ϕ Duplex plug receptacle

ϕ Split duplex receptacle used above kitchen counter

COUNTING OUTLETS ON A CIRCUIT

(1). A single or duplex receptacle counts as 1 outlet.

(2). A junction box does not count as an outlet (it is not one of the 12 outlets permitted on a circuit).

(3). A switch box does not count as an outlet (it is not one of the 12 outlets permitted on a circuit).

(4). Track lighting - each 1.5 m or fraction thereof of track counts as one outlet. For example a 2 m long lighting track would be counted as 2 outlets.

LIGHT OUTLETS REQUIRED

Rule 2-316 & 30-502 - The Electrical Code now requires at least one lighting outlet at, or in, the following locations:
• Each entrance - An exterior light fixture controlled by a wall switch located inside the house.
• Kitchen - A light fixture, controlled by a wall switch
• Bedrooms - A light fixture controlled by a wall switch or a wall receptacle controlled by a wall switch.
• Living room - A light fixture controlled by a wall switch or a wall receptacle controlled by a wall switch.
• Utility room - A light fixture controlled by a wall switch
• Dining room - A light fixture controlled by a wall switch
• Bathrooms - A light fixture controlled by a wall switch
• Washrooms - A light fixture controlled by a wall switch
• Vestibule - A light fixture controlled by a wall switch
• Hallways - A light fixture controlled by a wall switch
• Electrical Service Panel - One light fixture for operation and maintenance of that equipment.

STAIRWAY LIGHTING

Rule 30-504. All stairways must have illumination - there is no exception to this Rule. These rules say that it must be possible to control stairway lights from both the head and foot of the stairway. There is an exception. Where the stairway has only three or fewer risers, or in the case of a basement stairway leading to a dungeon like space which has no finished area and no egress other then the one stairway leading into it, one switch at the head of the stairway is acceptable.

The light outlet nearest the stairway must be wall switched at the head of the stairs.

Watch this one! The Rule says you must be able to turn off that stairway light from either the top or bottom of the stairway. In the case of an open stairway leading to a living room, or other such area on a different level, the normal switch locations for general lighting in that area may not also meet the requirements for stairway lighting. Additional switching may need to be installed or separate lighting and control may need to be installed for the stairway. Note that illumination is required at all stairways but the special switching requirement applies only in cases where the stairway consists of four or more steps.

UNFINISHED BASEMENT LIGHTING

Rule 30-506 says that "A lighting outlet with fixture shall be provided for each 30 m^2 (323 sq. ft.) or fraction thereof of floor area in unfinished basements." Note that the rule does not say there must be a lighting fixture "in" each 30 m^2 (323 sq. ft.) area in the basement. Compute the total floor area of the basement then divide by 30 or by 323 to determine the minimum number of light outlets required by this rule. These light outlets should be spaced to provide reasonably uniform light distribution over the whole floor area. One more detail, the light fixture nearest the stairway must be controlled by the one switch at the head of the stairs as required above under "Stairway Lighting."

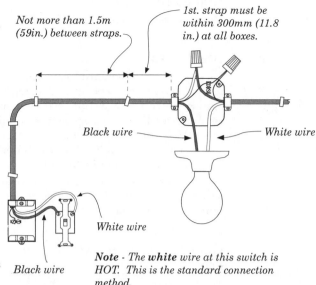

Not more than 1.5m (59in.) between straps.

1st. strap must be within 300mm (11.8 in.) at all boxes.

Black wire   *White wire*

STORAGE ROOMS

Rule 30-508 - This rule requires a lighting outlet with fixture in each storage room. The rule does not require this light outlet to be wall switched, but that is best.

White wire

Black wire

*Note - The **white** wire at this switch is HOT. This is the standard connection method.*

GARAGES AND CARPORTS

Rule 30-510 This rule requires a light outlet with fixture in each garage and carport. If the fixture is mounted above the car parking space it must be wall switched. If it is mounted off to the side it may be a switched lamp holder type provided that it is within reach of a person of average height.

In the case where the entrance way lighting fixture also provides light for the carport area, the rule says we do not need to install any additional lighting in the carport.

SIMPLE LIGHT OUTLET

The standard light outlet consists of a loomex cable run into an octagon or round outlet box set flush with the ceiling finish.

RECESSED LIGHT OUTLET

Rules 30-900, 30-910 - The wide variety of recessed light fixtures available today makes it difficult to be specific on installation instructions. The best advice, as always, is follow the manufacturer's instructions. Some IC and non IC fixtures are similar in appearance therefore check each fixture to be certain you are installing the correct fixtures. Do not deviate from the manufacturer's instructions. Basically there are three different types of recessed light fixtures available:

All connection leads must be long enough to reach drown through the fixture opening for future maintenance purposes.

IC TYPE

This recessed light fixture is certified for direct contact (blanketing), with building insulation and may be in contact with wood support members, Rule 30-906.

This is a specially designed, thermally protected, recessed light fixture which is totally enclosed in a metal box. This fixture may be covered with building insulation.

Lamp Wattage Rating - At present fixtures certified for covering with building insulation are limited to 75 watt lamps with appropriate trim. In some cases a simple change in fixture trim will alter the permitted lamp rating - check with your supplier.

As noted above, this fixture is equipped with a built-in thermostat to shut itself off in case it overheats. Some homeowners have installed 150 watt lamps in these 75 watt fixtures. It works fine, for a few minutes until the higher wattage lamp overheats the fixture. At this point the thermostat shuts it off. When it cools it turns itself back on again. Obviously, it could not do this for very long. The thermostat is not designed for repeated on off switching, it would very soon fail. Make sure you use the correct size lamp in your fixture.

IC INHERENTLY PROTECTED RECESSED LIGHT FIXTURE

This means its construction and the special lamps used in this fixture are designed to limit the heat generated so that it does not require a thermal protective device normally found in the IC type fixture noted in item (1) above. These are generally low output fixtures intended for special effects. This fixture may be in contact with building insulation, Rule 30-906.

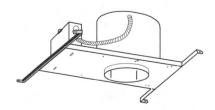

NON IC TYPE

This is the standard recessed light fixture approved for locations where there is no building insulation.

FIRE HAZARD

Improperly Installed Recessed Light Fixtures - A number of fires have been caused by incorrectly installed recessed light fixtures. There are two problem areas to watch out for when installing recessed light fixtures. First, make sure you have provided all the minimum clearance required in the manufacturer's instructions. Second, make sure all recessed light fixtures installed in insulated areas are properly certified for blanketing with building insulation. Look for the manufacturers label which clearly states the fixture is acceptable for blanketing with building insulation. Without this label the fixture should be rejected for that location. The Inspector is working to make your home safe. Because these fixtures can be a fire hazard you really want the Inspector to be very careful.

The insulation does exactly what it is supposed to do - it traps the heat in the fixture. When the fixture reaches the combustion temperature of the wood or paper next to it the result is charring and sometimes fire. Use only fixtures certified and marked approved for blanketing with building insulation.

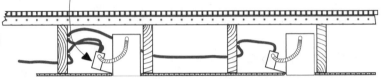

This box must be marked approved for use as a junction box, Rule 30-910(8), because the cable continues on to the next light.

CONNECTION OR TERMINAL BOX

The connection box on a recessed light fixture may be used only for the supply conductors to this one light fixture. You may not use it as a junction box for any other loads unless the fixture connection box is approved for that purpose. You will find many of these fixtures are equipped with boxes which are certified for use as a junction box. If they are acceptable for use as a junction box they will be so marked.

The illustration above shows a second floor above the recessed light fixture. In this case there is no requirement for building insulation and therefore, standard recessed light fixtures may be used. However, these fixture can still be a fire hazard if the minimum 1/2 in. clearance from wood is not maintained all around and on top of the fixture. The only points where the 1/2 in. clearance is not required is where the ceiling finish material butts up against the fixture and at the support points around the lower edge.

COMBINATION HEAT LAMP AND FAN FIXTURE

Unless the fixture is marked certified for blanketing with building insulation it should be regarded as not approved for covering.

Personal Opinion - If these fixtures are not CSA certified for direct covering with building insulation, it may be dangerous to cover them. This particular fixture is usually equipped with a 250 watt heat lamp which generates a great deal of heat. As long as the fan continues to operate it will tend to keep the internal temperature of the fixture to a safe level. In the event the fan fails

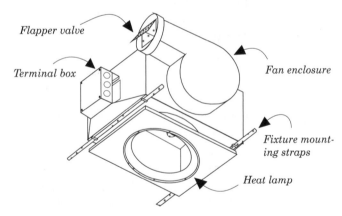

Flapper valve

Terminal box

Fan enclosure

Fixture mounting straps

Heat lamp

to operate, for whatever reason, fixture temperature may rise above the safe level. The ducting from this fan to outdoors will provide some natural ventilation provided there is no automatically operated flapper valve in this duct which is closed when the fan is not operating. It would also be necessary for the ducting to be adequately inclined to provide a chimney effect for heated air to move away from the fixture. Finally the Building Code requirements for very well sealed houses may stifle air movement to a point where you would have a 250 watt heater in the ceiling covered with R40 insulation. If the fan fails, who knows how long it would take start a fire.

FLUORESCENT LIGHT OUTLETS

Rules 12-506(1), 12-3002(6), 30-310(3) - Loomex cable may be run directly into a fluorescent light fixture as shown. Where fluorescent fixtures are mounted end to end as in valance or cornice lighting, the loomex cable should enter only the first fixture. The interconnecting wires between fixtures must be R90 or better.

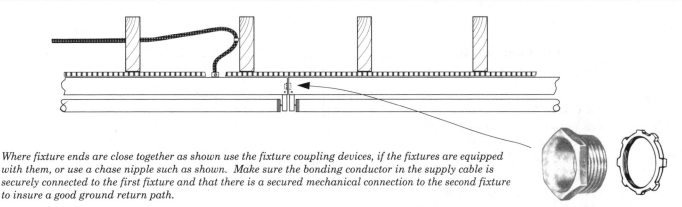

Where fixture ends are close together as shown use the fixture coupling devices, if the fixtures are equipped with them, or use a chase nipple such as shown. Make sure the bonding conductor in the supply cable is securely connected to the first fixture and that there is a secured mechanical connection to the second fixture to insure a good ground return path.

That threaded thing with a locknut next to it in the illustration, is called a chase bushing. Punch out the knockout holes in the end of the fixtures where they join. The fixtures are then fastened to the ceiling with their ends as close together as possible. The chase nipple is now inserted through the two knockouts and the locknut is used to bring the two fixture end plates together for grounding. This chase bushing also provides a smooth throat to protect the fixture supply conductors.

OUTLET BOX NOT REQUIRED

As shown, the loomex cable is run directly into the fixture. An outlet box is not required provided the cable used is NMD7 or NMD90 and not more than two cables enter any fixture.

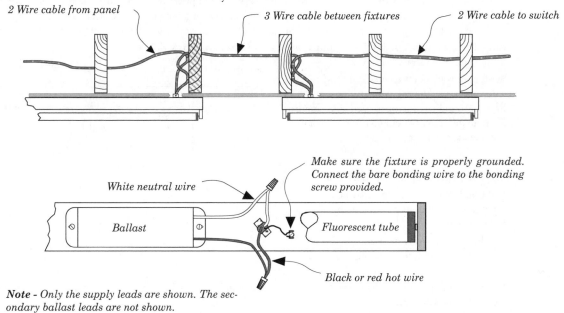

Note - Only the supply leads are shown. The secondary ballast leads are not shown.

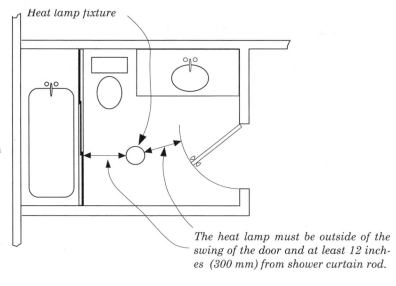

Heat lamp fixture

BATHROOM LIGHT OUTLETS

Rule 62-110(1)(b)

HEAT LAMPS

Like any other recessed light fixture, heat lamps can be a very real fire hazard if improperly installed. Care should be taken to:

(1). Locate the Heat Lamp fixture away from the door so that it cannot radiate heat directly onto the upper edge of the door when it is in the open position. This applies to the shower stall doors or curtain rod as well as the bathroom entry door. The rule does not specify a distance but some Inspection Authorities require at least 30 cm (12 in.), horizontal measurement, between the edge of the fixture and a shower rod or a door in any position.

The heat lamp must be outside of the swing of the door and at least 12 inches (300 mm) from shower curtain rod.

The reason for all this is to eliminate a possible fire hazard. The upper edge of the door may be too close to the fixture and could become overheated if the lamp was inadvertently left on for a long time; it is, after all, a heat lamp. The shower rod could be used to hang towels and clothing. These could become overheated if they were directly under the heat lamp.

Combination heat lamp/fan fixture for use in a bathroom is illustrated on p. 62.

Swag Lamps in a Bathroom - Rule 10-514(2) - Be sure to use the correct fixture - it must have a ground conductor to each chain hung lamp holder. Do not depend on the chain to provide adequate grounding.

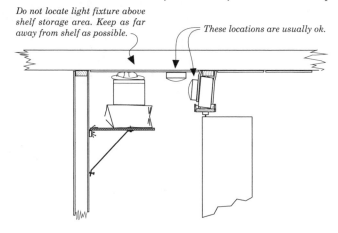

Do not locate light fixture above shelf storage area. Keep as far away from shelf as possible.

These locations are usually ok.

CLOTHES CLOSET LIGHT OUTLETS

Light outlet boxes in closets may be in the ceiling or on the front wall above the door, Rule 30-204(1).

The Rule requires an enclosed light fixture. A bare lamp type fixture is no longer acceptable in a closet.

Do not locate this light outlet above the shelf where it could be a fire hazard. Use great care in locating this fixture. It could be left on for days at a time.

SMOKE ALARMS & CARBON MONOXIDE ALARMS

Rule 32-110 - The Electrical Code requires alarm systems or carbon monoxide alarms; the Ontario Provincial Building Code may require both, and the National Building Code now requires one or more smoke alarms, and one or more carbon monoxide alarm devices in each residential dwelling. Please check with your local Building Inspector.

The number of smoke alarm units and carbon monoxide alarm units required for your house, and the location of these devices is strictly controlled by the local Building Inspector or Fire Officer, make sure they are satisfied.

Outlet Box Required - Use a standard light outlet box mounted as for a light outlet. This box may also be used as a junction box from which to serve other loads.

Supply Circuit - The smoke alarm and the carbon monoxide alarm should be supplied with a 15 amp circuit used for general lighting and plug outlets so that it becomes obvious when there is a power failure.

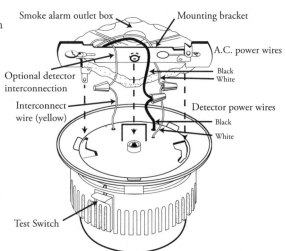

Smoke alarm outlet box

Mounting bracket

A.C. power wires

Optional detector interconnection

Black
White

Interconnect wire (yellow)

Detector power wires

Black
White

Test Switch

Do not supply these alarm units with a separate circuit used for other loads. Do not supply these alarm devices with a circuit which is either GFCI or AFCI protected. Do not switch this outlet. It must not be possible to turn these alarm systems off except using the breaker in the panel.

Carbon Monoxide Alarm

Choose carefully - the circuit you will use for this load, not because it is a big load, which it is not, but because it is very important to notice immediately when that circuit is open for any reason whatever. Because, when that circuit is open you are not protected. The circuit used for kitchen lighting may be the best choice because, as we grow older we get to like the kitchen even more than the bedroom and we would notice it more readily if the kitchen lights went out.

TWO OR MORE SMOKE ALARMS

Where two or more smoke alarms are installed, the Building Code says they must all operate together - that is, if one alarm is activated to sound an alarm the others must all be connected together so that they all automatically sound the alarm together. Units designed for line voltage (120 volts) require a 2-wire #14 supply cable to the first unit, then 3-wire from there to all the other units. The third conductor in these cables is for the signal circuit so that all units can sound the alarm together, the other two conductors are required to supply power to the second and third units. This type is in common use today.

OVERHEAD ROTARY FANS

These fans are becoming more and more popular. They not only look smart they also serve a useful purpose by moving the heated air downward to the floor level. - Some things to watch for:

- Look for a certification label. Never purchase any electrical appliance unless it is clearly marked with a CSA or one of the other certification labels described on p. 2. This is your protection and assurance that the device has been checked against a good standard.

- Look for mounting instructions. - Each fan has a caution marking which gives the minimum mounting height above floor required for that particular fan. This marking will look something like this: "Caution: Mount with the lowest mounting parts at least 8 ft. above floor or grade level."

Minimum elevation above floor level is 8 ft.; except that in some cases it may be as low as 7 ft.

- Use a special mounting box to support the fan. -Some fans will not turn as fast at maximum speed or the fan blades are designed so they are less hazardous to anyone coming in accidental contact with the blades. These fans will also have caution markings similar to the words given above except that in this case the minimum mounting height will be 7 ft. instead of 8 ft. In this case the fan blades may be as low as 7 ft. above the floor. Also see, pendant fan mounting requirements discussed under "Light Outlet Boxes" on p. 59

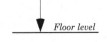
Floor level

- Where a ceiling outlet box is marked for fan support, it must be securely attached directly to the building structure, or attached by a bar hanger, which is securely attached to the building structure.

NEAR STAIRWAYS, BALCONIES, AND SUCH LIKE

The fan blades may not be within reach of a person standing on a stairway, a landing or a balcony. If the blades are within reach the minimum height given on the caution notice must be measured from the level the person is standing on. Check this detail carefully in the rough wiring stage - it is very difficult to change the location of the supply outlet later.

CSA has issued a caution regarding these fans. It warned that if the blades are not properly installed they may work loose and fall to the floor. Anyone in the path of such a flying blade could be seriously hurt.

CIRCUIT REQUIRED

The rotary fan may be supplied with any lighting circuit which has only 11 or fewer outlets. The fan outlet, though it is a small load, counts as one outlet.

VAPOUR BARRIER

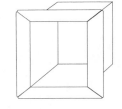

This is not an Electrical Code requirement, it is part of the Building Code. This means that Electrical Inspectors do not ask for these boxes, but they are still very concerned how they are installed. Where the Building Code requires a vapour barrier on a wall or ceiling, (normally this is required only on insulated walls or ceilings) all electrical boxes located in that wall or ceiling must also be enclosed in a vapour barrier. This vapour barrier for an electrical box is like a shroud. It consists of a separate plastic or fiberglass box, which has a wide gasketted flange to allow contact, and seal with the vapour barrier sheet that is normally fixed to the joists or studs. The electrical outlet box is placed inside this shroud, then together, they are fixed to a stud. Some box manufacturers produce what they call airtight boxes. These are equipped with a wide flange and seals where cables must enter the box. Whichever kind of vapour barrier you use, the Building Code requires this barrier at:

- Each light outlet box
- Each light fixture if it is recessed into the wall or ceiling
- Each switch outlet box
- Each plug outlet box
- Each outlet box for other purposes such as a smoke alarm J-box etc.

One more thing: don't forget to apply a sealing compound around each cable entering the outlet box. The whole idea is to make the house fully air tight. There must not be any unauthorized fresh air entering the house.

A recessed light fixture located in an insulated ceiling must also be equipped with a vapour barrier. These must enclose the whole fixture thus requiring a rather large box as shown at left.

Don't forget this detail. It's easier to install these to begin with then later, after the rejection.

The two boxes shown below are equipped with sealing flanges. These are readily available in local building supply stores.

There is a rumor that says this whole exercise is designed to make our houses air tight so that the stale air inside cannot get out to pollute the atmosphere, it's probably true.

17B. SWITCH OUTLET BOXES

VAPOUR BARRIERS FOR SWITCH OUTLET BOXES

The Building Code requires boxes to be surrounded with a moisture resistant barrier if they are located in a wall or ceiling which is required to have a vapour barrier. See p. 65 for details.

CABLE CONNECTORS

Switch, light, plug and junction boxes normally have four separate cable entry holes for cables and each hole is equipped with a cable connector. Each of these cable entry holes is certified for one only cable entry, not two cables. Each hole may contain one 2-wire cable or one 3-wire cable. Plan, and arrange your circuit runs, so that there is no need to enter a connector with two cables.

HEIGHT OF SWITCHES

The rules do not specify a required height for wall mounted light switches. They may be located at any convenient height - usually they are set at approximately 1.2 m (approx. 48 in.) to the lower edge of the box.

BATHROOM LIGHT SWITCH

Rule 30-320(3) -The light switch in a bathroom must be located at least 1 m (39.4 in.) distance between the bathtub or shower stall. This means that switches controlling these lights may be located inside the bathroom provided they are at least 39.4 in. from the a bathtub or shower stall. Where this distance is not practicable it can be reduced even further but it must not be less than 50 cm (19.69 in.). In any case, where this distance is less than 1 m (39.4 in.), the lighting circuit must be GFCI protected.

Bathroom Thermostat for electric heaters - Rule 62-202 requires the thermostat control for an electric heater in a bathroom to be at least 1 m (39.4 in.) horizontal distance from the outside face of a bathtub or shower stall. Where this simply isn't possible, the switch can be located not less than 50 cm (19.69 in.) from the bathtub or shower provided that it is GFCI protected. See also "Thermostats," p. 123.

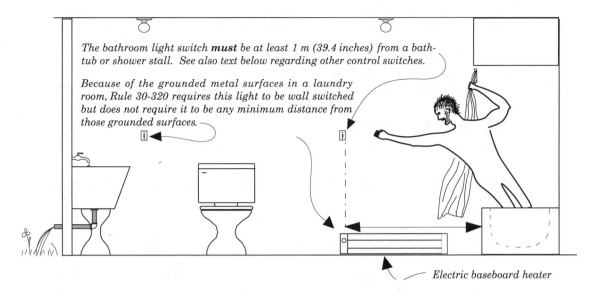

*The bathroom light switch **must** be at least 1 m (39.4 inches) from a bathtub or shower stall. See also text below regarding other control switches.*

Because of the grounded metal surfaces in a laundry room, Rule 30-320 requires this light to be wall switched but does not require it to be any minimum distance from those grounded surfaces.

Electric baseboard heater

Horizontal Measurement - The Code requires this to be a horizontal measurement from the switch to the nearest outside face of a shower stall, or tub, as illustrated above.

Other Switches & Controls Located in a Bathroom - The Code does not specifically say all controls and switches located in a bathroom must be 1 m (39.4 in.) from a bathtub or shower stall but they are all equally dangerous if they operate at the same voltage.

Heat Lamp Control Switch - The switch for the infrared heat lamp should (not must) be at least 39.4 in. (1 m) from the bathtub or shower stall.

Thermostat Control - Thermostats located on a wall or on a heating unit should (not must) be at least 39.4 in. (1 m) horizontal distance from the outside face of a tub or shower stall.

Bathroom Fan Switch - This switch is just as dangerous as the others listed above and should (not must) also be located at least 39.4 in. (1 m) from the outside face of a tub or shower stall.

Hydro massage bathtub Control - See p. 120 for details.

KITCHEN AND LAUNDRY ROOM LIGHT SWITCHES

Subrules 30-320(1)&(2) refer to lighting equipment located near grounded metal surfaces. It refers to laundry tubs, plumbing fixtures, metal pipes etc. installed in damp locations and it requires the lighting in such areas to be wall switched but does not require them to be GFCI protected.

This Rule does not mention switches for kitchen lighting or any other switches located near a kitchen sink. Such switches could be located on either side of the sink but should not be directly behind the sink or directly in front of the kitchen sink. These switches are also not required to be GFCI protected.

STAIRWAY LIGHT SWITCH

Rule 30-504 requires illumination at all stairwells; there is no exception. - Control switches for these light outlets are required at both head and foot of each stairway in all cases except where the stairway has only three or fewer risers and in the case of a basement stairway leading into a dungeon like space which has no finished area and no egress other than the one stairway leading into it.

Watch this one! The Rule says you must be able to turn off that stairway light from either the top or bottom of the stairway. This may not be possible with the area lighting switch in the case of open stairways leading into lower level rooms or areas. See p. 60 for details. The rule refers to short stairways consisting of from one to three risers only. These must be lighted, but special lighting and specific control is not required. General area lighting and it's control is normally acceptable to this Rule.

Four or more risers - The Rule also refers to stairways consisting of four or more risers. These stairways must be specifically lighted and controlled with a 3-way switch located at the head and another at the foot of the stairs. There is an exception in the case of a basement stairway which leads to a dungeon like space which has no finished area and no egress other than the one stairway leading into it. One switch at the head of such a basement stairway is acceptable. See p. 60 for details.

CONNECTION OF SWITCHES

Rules 4-034(2), 30-602 - Switch connections shall be made so that there is a white wire and a black wire to the fixture. To do this the connection should be made as follows: The white wire in the supply cable shall connect directly to the screw shell (the silver terminal) in the lamp holder. The black wire from the switch connects to the center (gold) terminal in the lamp holder. In this way the fixture has a black and a white wire supplying it. It also has the black wire connected so that the screw base of the bulb cannot become energized. This is very important.

The following drawings illustrate acceptable connection arrangements in switch and light outlet boxes.

SWITCHING ARRANGEMENTS

See "Actual Box Fill Permitted Table" on p. 105 in conjunction with this section for additional switching scenarios.

POWER ENTERING LIGHT OUTLET BOX FIRST (SIMPLEST SWITCHING ARRANGEMENT)

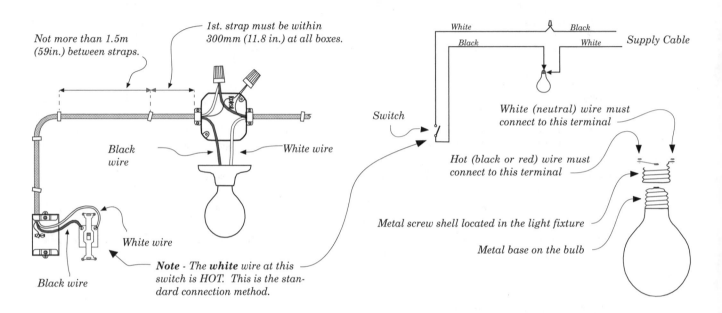

The white wire, in the case of switch legs, is hot as shown. This is the normal Code connection. This white wire does not need to be wrapped with black tape or be painted black,. Leave it as is.

Make sure that you always connect the neutral wire (the white wire) in the supply cable to the screw shell in the fixture. Your life could depend on this detail. See "Lighting Fixtures," p. 110, and "Conductor Joints and Splices," p. 101.

POWER ENTERING SWITCH OUTLET BOX FIRST

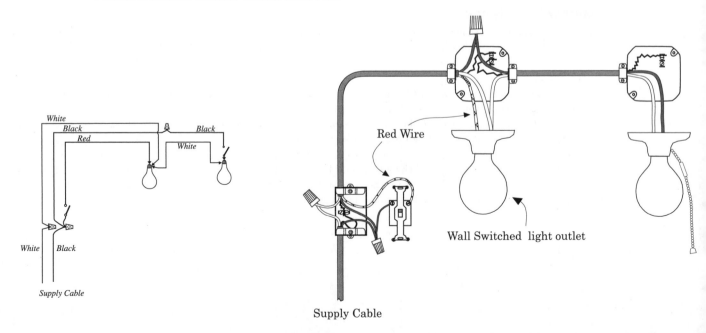

POWER FEEDING THROUGH SWITCH BOX TO PLUG OUTLET

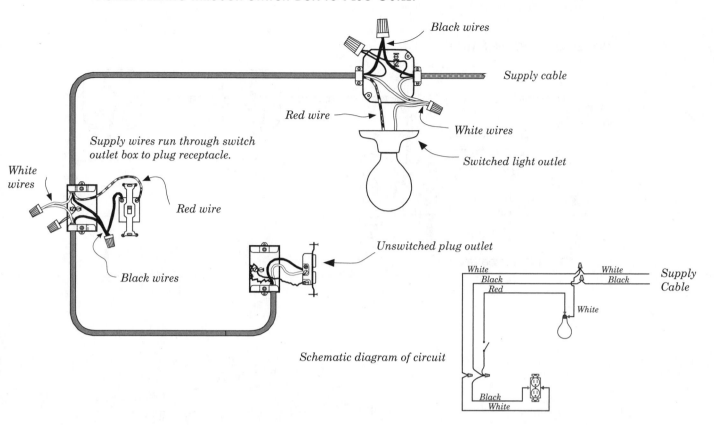

Black wires

Supply cable

Red wire

White wires

Supply wires run through switch outlet box to plug receptacle.

Switched light outlet

White wires

Red wire

Black wires

Unswitched plug outlet

Schematic diagram of circuit

White · Black · Red · White · Black · Supply Cable · White · Black · White

3-WAY SWITCHES

BOTH SWITCHES ON THE SAME SIDE OF THE LIGHT (SIMPLE ARRANGEMENT)

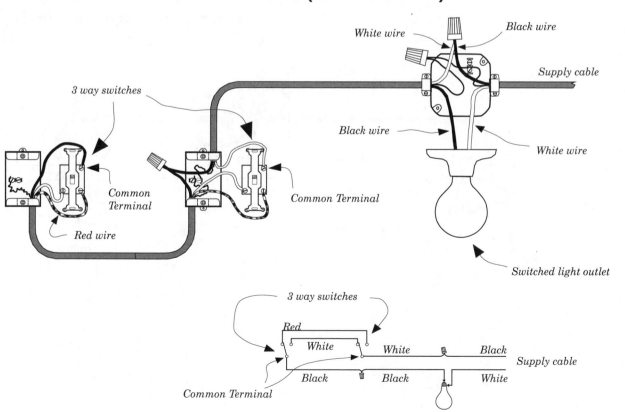

White wire   *Black wire*

Supply cable

3 way switches

Black wire

White wire

Common Terminal

Common Terminal

Red wire

Switched light outlet

3 way switches

Red · White · White · Black · Supply cable · Black · Black · White

Common Terminal

Bond wires must also be properly spliced and connected in every outlet box. To avoid confusion these bond wire connections are not always shown. See "Bonding Light Fixtures," p. 111.

POWER ENTERING THE SWITCH

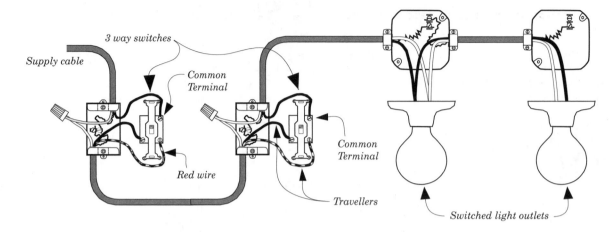

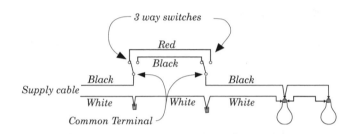

LIGHT BETWEEN SWITCHES (POWER ENTERING THE SWITCH)

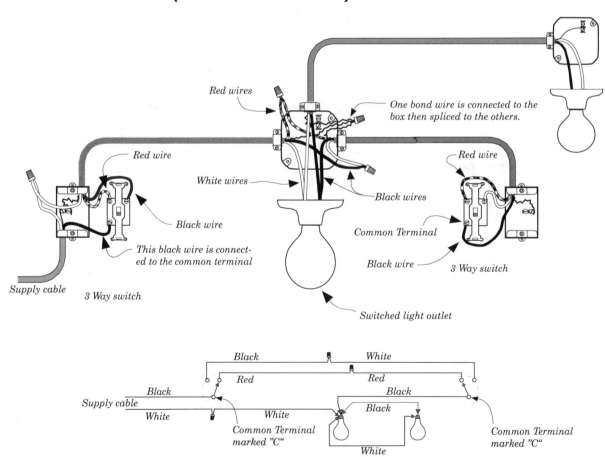

LIGHT OUTLET BETWEEN SWITCHES (POWER ENTERING THE LIGHT OUTLET BOX)

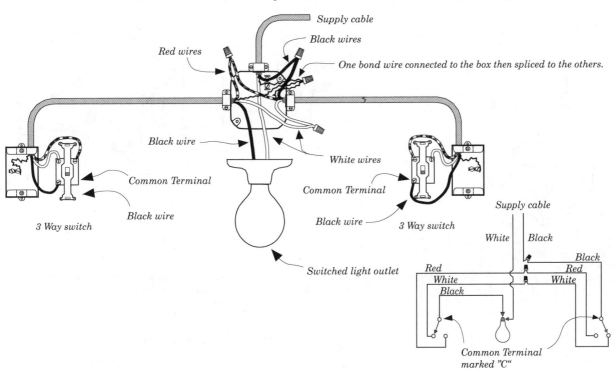

Supply cable

Black wires

Red wires

One bond wire connected to the box then spliced to the others.

Black wire

White wires

Common Terminal

Common Terminal

Black wire

Black wire

3 Way switch

3 Way switch

Switched light outlet

Supply cable

White   Black

Red

Black

White

Red

Black

White

Common Terminal
marked "C"

4-WAY SWITCH CONTROL (FOR TWO LIGHT OUTLETS)

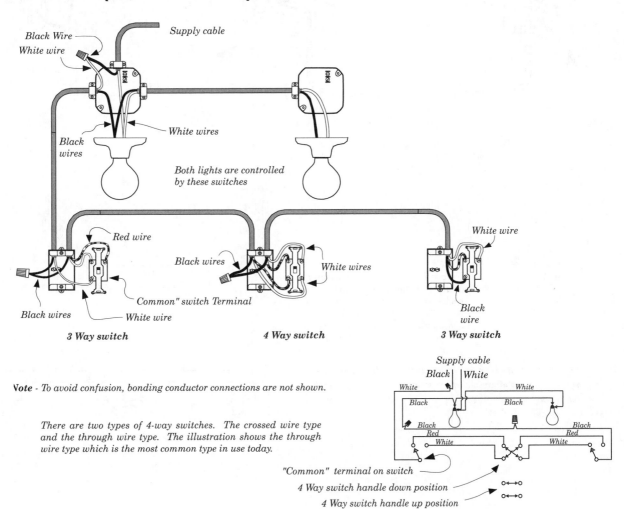

Black Wire
White wire

Supply cable

White wires

Black
wires

Both lights are controlled
by these switches

Red wire

White wire

Black wires

White wires

Black
wire

Black wires

Common" switch Terminal

White wire

3 Way switch   **4 Way switch**   **3 Way switch**

Note - To avoid confusion, bonding conductor connections are not shown.

There are two types of 4-way switches. The crossed wire type
and the through wire type. The illustration shows the through
wire type which is the most common type in use today.

Supply cable
Black   White

White   White

Black   Black

Black   Black
Red   Red
White   White

"Common" terminal on switch

4 Way switch handle down position

4 Way switch handle up position

2-GANG SWITCH BOX WITH TWO SWITCHED PLUGS

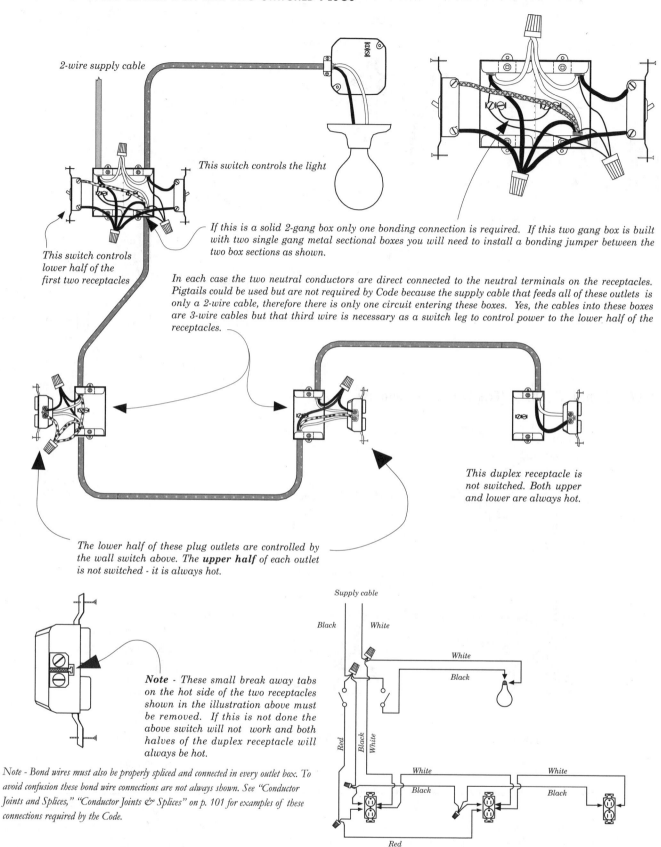

2-wire supply cable

This switch controls the light

This switch controls lower half of the first two receptacles

If this is a solid 2-gang box only one bonding connection is required. If this two gang box is built with two single gang metal sectional boxes you will need to install a bonding jumper between the two box sections as shown.

In each case the two neutral conductors are direct connected to the neutral terminals on the receptacles. Pigtails could be used but are not required by Code because the supply cable that feeds all of these outlets is only a 2-wire cable, therefore there is only one circuit entering these boxes. Yes, the cables into these boxes are 3-wire cables but that third wire is necessary as a switch leg to control power to the lower half of the receptacles.

This duplex receptacle is not switched. Both upper and lower are always hot.

*The lower half of these plug outlets are controlled by the wall switch above. The **upper half** of each outlet is not switched - it is always hot.*

Note - *These small break away tabs on the hot side of the two receptacles shown in the illustration above must be removed. If this is not done the above switch will not work and both halves of the duplex receptacle will always be hot.*

Note - Bond wires must also be properly spliced and connected in every outlet box. To avoid confusion these bond wire connections are not always shown. See "Conductor Joints and Splices," "Conductor Joints & Splices" on p. 101 for examples of these connections required by the Code.

Supply cable

Black   White

White

Black

Red   Black   White

White

Black

White

Black

Red

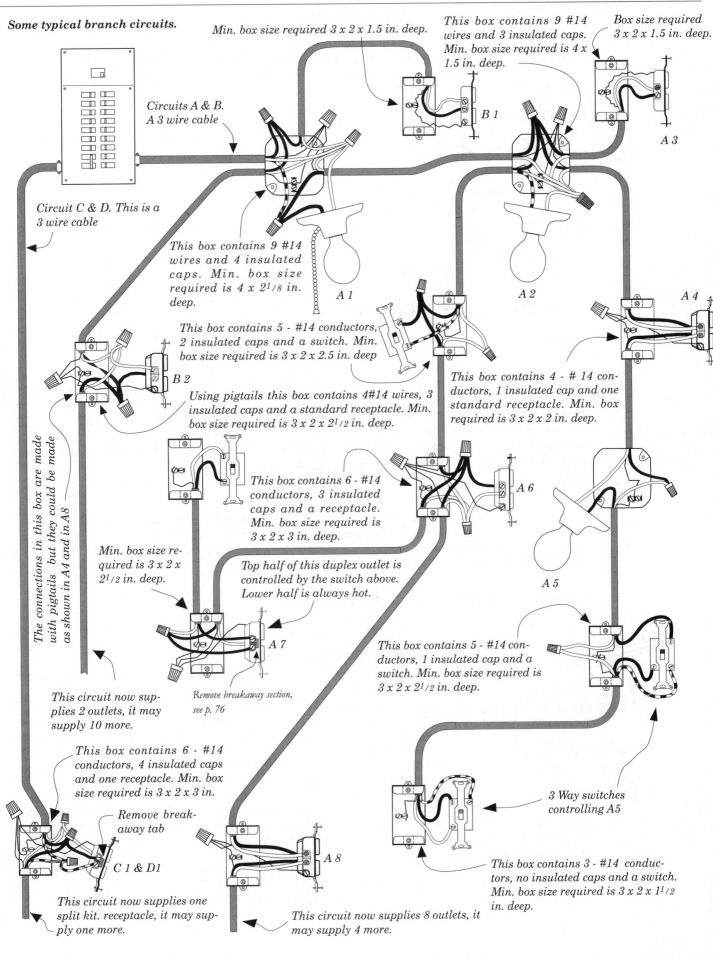

Some typical branch circuits.

Min. box size required 3 x 2 x 1.5 in. deep.

This box contains 9 #14 wires and 3 insulated caps. Min. box size required is 4 x 1.5 in. deep.

Box size required 3 x 2 x 1.5 in. deep.

Circuits A & B. A 3 wire cable

B 1

A 3

Circuit C & D. This is a 3 wire cable

This box contains 9 #14 wires and 4 insulated caps. Min. box size required is 4 x 2 1/8 in. deep.

A 1

A 2

A 4

This box contains 5 - #14 conductors, 2 insulated caps and a switch. Min. box size required is 3 x 2 x 2.5 in. deep

B 2

Using pigtails this box contains 4#14 wires, 3 insulated caps and a standard receptacle. Min. box size required is 3 x 2 x 2 1/2 in. deep.

This box contains 4 - # 14 conductors, 1 insulated cap and one standard receptacle. Min. box required is 3 x 2 x 2 in. deep.

The connections in this box are made with pigtails but they could be made as shown in A4 and in A8

This box contains 6 - #14 conductors, 3 insulated caps and a receptacle. Min. box size required is 3 x 2 x 3 in. deep.

A 6

A 5

Min. box size required is 3 x 2 x 2 1/2 in. deep.

Top half of this duplex outlet is controlled by the switch above. Lower half is always hot.

A 7

Remove breakaway section, see p. 76

This circuit now supplies 2 outlets, it may supply 10 more.

This box contains 5 - #14 conductors, 1 insulated cap and a switch. Min. box size required is 3 x 2 x 2 1/2 in. deep.

This box contains 6 - #14 conductors, 4 insulated caps and one receptacle. Min. box size required is 3 x 2 x 3 in.

Remove breakaway tab

3 Way switches controlling A5

C 1 & D1

A 8

This box contains 3 - #14 conductors, no insulated caps and a switch. Min. box size required is 3 x 2 x 1 1/2 in. deep.

This circuit now supplies one split kit. receptacle, it may supply one more.

This circuit now supplies 8 outlets, it may supply 4 more.

Circuits A & B are supplied with a 3-wire cable to the first light outlet where it splits into 2 - 2-wire cables. Each of the outlets, supplied by these cables, is connected to only one hot wire and the neutral, therefore, the circuit breakers supplying this 3-wire cable do not need to be equipped with a tie-bar. The 3-wire cable on the left supplies kitchen counter outlets. Each of these is connected to both hot wires and the neutral, therefore, this cable must be supplied with either a two pole circuit breaker or two single pole breakers with their operating handles tied together with a tie-bar. See p. 37 for very important details on 3-wire cable connections to breakers in the panel.

18.   PLUG OUTLETS

RULES 26-700 & 26-702

OUTLETS PER CIRCUIT

A maximum of 12 outlets may be connected to a circuit. This may consist of 12 light outlets or 12 plug outlets (not appliance plugs, see "Kitchen Counter Plug Outlets," p. 82) or any combination of light and plug outlets mixed, as long as their total number does not exceed 12 outlets.

It is better to have the load consist of a mixture of lights and plugs. This gives better load diversity on the circuit and less chance of a complete blackout in case of circuit failure.

VAPOUR BARRIER FOR A PLUG OUTLET BOX

The Building Code requires boxes to be surrounded with a moisture resistant barrier if they are located in a wall or ceiling which is required to have a vapour barrier. See p. 65 for details.

HEIGHT OF PLUG OUTLETS

The rules do not specify any definite height for plug outlets. They may be at any convenient height. Usually they are placed at approximately 30 cm (approx. 12 in.) to lower edge of the outlet box in the living room, dining room, bedrooms and hallways etc.

Horizontal or Vertical - may be either way but if you want a professional looking job, install them all in the vertical plane. Besides, they look even more handsome that way.

CABLE CONNECTORS

Switch, light, plug and junction boxes normally have four separate cable entry holes for cables and each hole is equipped with a cable connector. Normally, each of these cable entry holes is certified for only one cable entry, not two cables. Each hole may contain one 2-wire cable or one 3-wire cable. Plan, and arrange your circuit runs, so that there is no need to enter any single connector with two cables.

RECEPTACLES - FAULT PROTECTION - AFCI AND GFCI

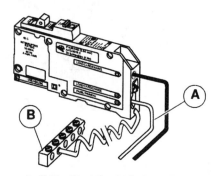

Arc-D-Tect Circuit Breaker by Square D

ARC-FAULT CIRCUIT INTERRUPTERS (AFCI)

An arc-fault circuit interrupter (AFCI) is a circuit breaker that can distinguish between a normal heavy load and an arcing short circuit. It provides protection against arcs that cause fires.

The 22nd Code edition made AFCIs mandatory for branch circuits suppling receptacles in bedrooms. The 23rd edition has expanded the requirement for AFCIs. It makes AFCIs mandatory for branch circuits supplying almost all receptacles in dwellings. Exceptions are the bathroom and washroom and many kitchen receptacles. The kitchen receptacles named as excluded from the requirement for AFCI are those for the refrigerator and some outlets along the wall at counter work surfaces, on kitchen islands, and on kitchen peninsulas. References: Subrules 26-724(f) and 26-712(d)(i), (iii), (iv), and (v).

The reason for requiring more AFCIs is that a standard circuit breaker could fail to open the circuit under certain short circuit conditions. A short circuit with massive current flow will normally trip a standard circuit breaker very quickly but a short circuit in which certain technical features limit the current that can flow, may not be recognized by such a circuit breaker. It may continue to feed the fault until enough heat is generated to ignite the surrounding combustibles such as wood or paper. Remember this is not a GFCI breaker and it cannot protect as a GFCI breaker. Because they protect in different ways, it's perfectly acceptable to have GFCI receptacles on AFCI circuits.

The illustration above shows an arc-fault circuit breaker manufactured by Square D Company. Its connection is similar to that of a GFCI breaker in that the lead from the breaker must terminate in the neutral bus as shown at (B) and both the black and the white conductor in the loomex cable must connect to the breaker as shown at (A).

This breaker cannot be used with three wire cables, only two wire cables may be used. Three wire cable used in three-way switch control is okay. This circuit breaker may also be used to protect lights and plug circuits in other parts of the house but it must be used to supply all the plug receptacles in each of the bedrooms.

Subrules 26-722(f) & (g) state that where AFCI protection is required it must be in the circuit breaker located in the panel so that it can protect the whole branch circuit. AFCI protection, if located in a plug receptacle, cannot protect the branch circuit conductors and therefore is not acceptable. GFCI protection is acceptable at the receptacle level and can operate on an AFCI protected circuit.

GROUND-FAULT CIRCUIT INTERRUPTERS (GFCI)

GFCI Protection - This device measures very carefully, the current flowing toward the load in the black wire, then compares that amount of current with the current flowing back to the panel in the white wire. If there is a difference it means that some current is flowing back to the panel through some other unauthorized route. That unauthorized route may be thorough a person holding a defective electrical device and the current is flowing through that person to ground. If this current is greater than 5 milliamperes, 5/1,000 of an ampere the GFCI will open the circuit. Thomas Edison never had one of these. He took all manner of risks so that after dark, we would still be able to see.

GFCI protection may be in the circuit breaker in the panel or in the plug receptacle.

WHAT'S A GFCI FOR? WHAT DOES IT DO? -IT SAVES LIVES!

When an electrical appliance is faulty the appliance itself may become energized. Anyone holding such an appliance could become part of that supply circuit to ground. They would become an electrical conductor and current would flow through the person to ground. Only a very small current is needed to kill a human being. The graph indicates the enormous difference between the small amount needed to kill a person and the large amount available in every 15 amp circuit in the house.

15 amp circuit.

This range is fatal.

50 mA range. This amount of current flowing through a person may cause fibrilation and death.

18 mA range. This amount of current flowing through a person will usually affect his respiration.

6 - 9 mA range. This is called the let go range. A person can usually let go even with this amount of current flowing through his body.

6 mA range. This is the GFCI-A breaker tripping threshold.

1 ma range. The perception range.

The graph also indicates the very low current that a GFCI will pass to ground. It is designed to trip at a maximum 5 m.a. which is only 0.005 ampere. This means that a person holding a faulty electrical appliance, such as an electric lawn mower or an electric drill, could become an electrical conductor to ground and they could get an electrical shock but the GFCI type circuit breaker would open the circuit before the current reached a dangerous level.

It would be very expensive to protect all the circuits in the house with these special circuit breakers. What's more, it is not necessary. The code requires only certain outlets to be protected with this special breaker. As indicated above, outdoor plug outlets are among those outlets which must be protected with this special circuit breaker.

Test required - To provide some measure of assurance that the GFCI protection is still faithfully protecting us, we are asked to test each GFCI device every month, and to record the test. If any one of them fails the test it should be removed from service immediately - do not use a failed safety device.

Test record - The test procedure information that comes with a GFCI protected circuit breaker or receptacle is important.

Manufacturers usually suggest that this safety device be tested each month. This is a simple procedure when the GFCI protection is in the receptacle but a little more difficult when the test procedure can only be completed with the circuit breaker in the service panel. Keep a written record of your tests on, or near, the service panel. It will serve as a reminder to test the circuit breaker on a regular basis. If possible, do the test under load with a shaver or a portable light fixture. This will confirm the circuit is actually opened in the test. If at any time this breaker or receptacle fails a test it should not be used. It should be replaced; do not use a failed safety device. If at any time this breaker fails a test it should be turned off until it can be replaced; do not allow it to be used.

RECEPTACLES - TYPE, USE, PURPOSE

Duplex receptacles - All 15 Amp and 20 amp plug receptacles installed both inside and outside of a single family house must be duplex type. Single receptacles are acceptable only for dedicated circuits for specific single use plugs, such as a microwave, a refrigerator, or a freezer, etc.

Standard receptacles - this is the basic receptacle. It is not tamper resistant and it is not usable on a split circuit. Very few of these are permitted in house wiring now under the new Code. These receptacles may be used only as follows:

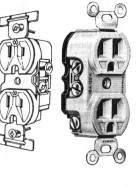

- For plug outlets dedicated for a microwave, a refrigerator or a freezer.
- For some kitchen counter plugs.
- For plug outlets in the attic.
- For plug outlets in the crawl space.

Otherwise, do not use these standard plugs.

Tamper Resistant (TR) receptacles - These look just like a standard duplex receptacle except that somewhere on the front face it will be marked to show that it is tamper resistant. The first detail to note is that these new receptacles are not tamper proof, they are only tamper resistant. After-all we must not make it too difficult, we may need to ask the little ones for help to operate these things.

Where are these required - These tamper resistant receptacles are required in all child care facilities and also throughout residential dwellings at each plug outlet that would normally be within reach of a small child. This includes both inside and outside the dwelling and for all plugs except those in the following locations:

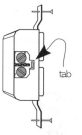

- Plug outlets dedicated for a microwave, a refrigerator or a freezer.
- Some kitchen counter plugs.
- Plug outlets in the attic.
- Plug outlets in the crawl space.

All other 15 amp & 20 amp plug outlets must be TR type.

Split receptacles - It has a small breakaway tab as shown. This receptacle is normally only used on the work kitchen counter and in that location it need not be Tamper Resistant.

T-Slot Receptacles - These are 20 Amp receptacles often used for kitchen counter plugs. These must be wir p. 82.

NUMBER OF PLUGS AND LOCATION

Living room
Family room
Rec. room
Bedroom
Den
Study

Rule 26-712(a)(c) require plug outlets in these rooms to be located so that it is not possible for an electrical appliance to be more than 1.8 m (approx. 6 ft.) from a plug outlet when it is located anywhere along the wall.

Note - This measurement is not a radius - you must measure into corners as shown in the illustration. This is the strict interpretation of Subrule (c).

The illustration below shows where these plugs must be located. It can be used as a template to show where plug outlets are required in each of the rooms in the above list.

Rule 26-712(c) says do not count spaces occupied by:

- Doorways, and the area occupied by the door when fully open.
- Window - The space occupied by windows that extend to the floor need not be counted.
- Fireplaces
- Other permanently fixed installations that limit the use of wall space.

As indicated in the drawing below, isolated sections of wall less than 90 cm long and those sections occupied by any permanent items are not considered to be usable wall spaces and therefore do not require a plug outlet, Rule 26-712(c):

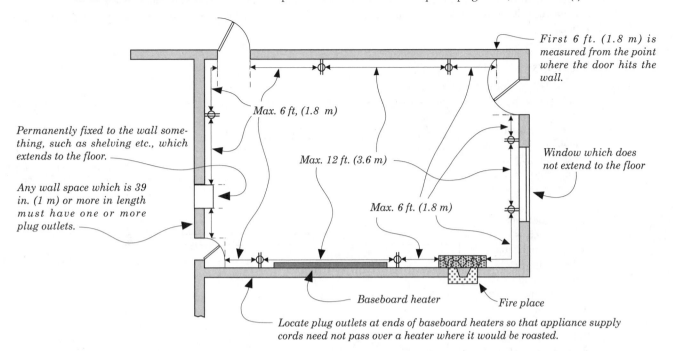

First 6 ft. (1.8 m) is measured from the point where the door hits the wall.

Max. 6 ft, (1.8 m)

Max. 12 ft. (3.6 m)

Max. 6 ft. (1.8 m)

Permanently fixed to the wall something, such as shelving etc., which extends to the floor.

Any wall space which is 39 in. (1 m) or more in length must have one or more plug outlets.

Window which does not extend to the floor

Baseboard heater

Fire place

Locate plug outlets at ends of baseboard heaters so that appliance supply cords need not pass over a heater where it would be roasted.

Electric & Other Types of Baseboard Heaters - are a permanent unit and they do limit the use of that wall space to some extent, however, appliances such as radios, T.V. swag lamps, etc. can be placed in that wall space and each of these requires power. The rule therefore, does require outlets in wall spaces occupied by baseboard heaters. These outlets should not be located above the heaters unless it cannot be avoided. Usually they can be located at the ends of the heaters so that electrical supply cords need not run over the heater and be roasted every time the heater comes on. See also Appendix B for Rule 26-712(a) in the Electrical Code for confirmation of this interpretation.

BEDROOMS

Note that bedrooms still require AFCIs, and most other rooms do, too, as of the 23rd Code edition.

ENTRANCE (FOYER)

Rule 26-712(a) - If this is a room, treat it as a living room. If it is like a hallway, apply the rule for hallways. If it is something in between, well, just put in the extra outlets and be done with it. Don't quibble over little things. Entrances (foyers) require AFCIs.

HALLWAYS

Rule 26-712(f) - Locate the plug outlet so that no point on the hallway floor is more than 4.5 m (15 ft) from a plug outlet without having to go through a doorway fitted with a door.

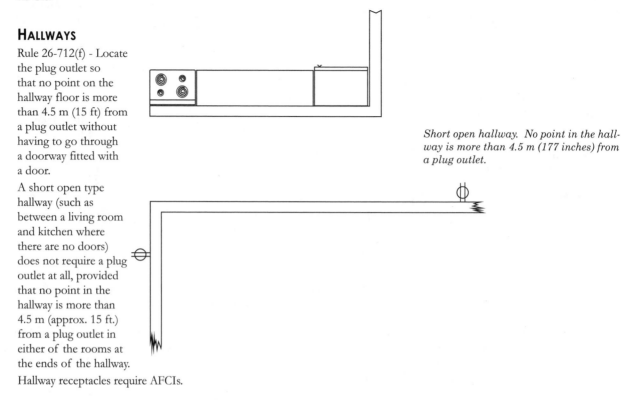

Short open hallway. No point in the hallway is more than 4.5 m (177 inches) from a plug outlet.

A short open type hallway (such as between a living room and kitchen where there are no doors) does not require a plug outlet at all, provided that no point in the hallway is more than 4.5 m (approx. 15 ft.) from a plug outlet in either of the rooms at the ends of the hallway.

Hallway receptacles require AFCIs.

BASEMENT WIRING

Rules 26-710(a)(b) & (e) and 26-712(a)(b)(c) - Finished Basement - The rules require the same number of outlets along the walls of a basement as for similar rooms upstairs. The walls of basement bedrooms, hallways, family rooms etc. that are finished must be wired as similar rooms on the main floor.

Basements receptacles require AFCIs.

UNFINISHED BASEMENT

First, a Definition of Unfinished - Rule 26-710(a) - The definition of a finished wall is perhaps a little strange and unlikely, but there needed to be clear guidance to make the Rules enforceable.

Unfinished basements require AFCIs where plugs (receptacles) are installed.

Basically, Rule 26-710 states that full wiring is not required in any wall which is substantially unfinished. If the wall finish material does not extend fully to the floor, i.e. if the lower 45 cm, or greater portion of the wall, is not finished with any kind of finishing material, that wall is considered unfinished. Such a wall is required to have only minimum wiring as described below. Building insulation and vapour barrier may be installed in all walls and it may extend to the floor. Building insulation is not finishing material.

Each wall or partition is considered separately. If the lower unfinished portion of the wall is less than 45 cm (17.7 in.) then the wall can be declared as a finished wall and therefore requires full normal wiring.

Minimum Basement Plug Outlets Required - Rule 26-710(e)(4) - requires only one plug outlet in an unfinished basement. If there are no walls to divide the basement into two or more rooms or areas the rule is satisfied with just one duplex plug outlet in the whole basement. See also below under "Laundry Plug in Basement."

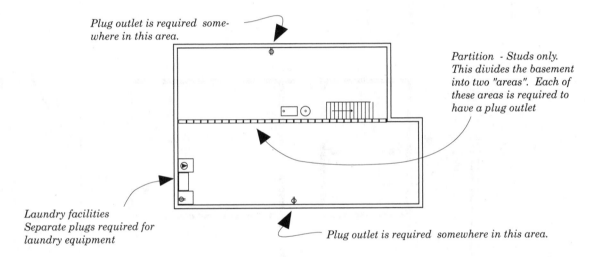

These walls are finished therefore full wiring is required.

If the wall finish extends downward to 17.72 inches above the floor line it is a finished wall.

If the unfinished lower portion of the wall is greater than 17.72 inches it is an unfinished wall.

17.72 inches

Note - Basement partitions - unfinished - studs only. If there are no partitions to divide the basement into two or more areas and the lower portion of the basement outside walls are not finished except as described above and there is no laundry facility in the basement, this rule is satisfied with just one duplex plug outlet in the whole basement.

Caution - Rule 26-710(e)(4) refers to an "area" not to a room. The rule says "at least one duplex receptacle shall be provided in any unfinished basement area." Unfinished partitions consisting of studs only, can and do divide the total basement floor space into two or more areas and each of these areas is required to have at least one duplex plug outlet. Note too that Rule 26-712(a) also refers to a "room or area" and says that both must be treated equally. There may be differences of opinion on this interpretation, therefore you should check with your local inspector before proceeding.

Laundry Plug in Basement. - Rule 26-710(e) - If laundry facilities are located in the basement then the plug outlet for the washer is in addition to the minimum plug outlets described above and it must be on its own circuit. AFCI protection is required for receptacles.

Plug outlet is required somewhere in this area.

Partition - Studs only. This divides the basement into two "areas". Each of these areas is required to have a plug outlet

Laundry facilities Separate plugs required for laundry equipment

Plug outlet is required somewhere in this area.

Fridge - Duplex receptacle on a separate circuit is required except that a recessed clock outlet may also be supplied with this circuit.

Counter Outlets - No point along the back edge of a kitchen counter may be more than 900 mm (35.5 in) from an outlet.
- 300 mm (11.8 in) or longer counter space requires a plug outlet.
- Adjacent plug outlets **may now be on the same circuit.**

Dining Area in Kitchen - At least one plug outlet is required - must be on a separate circuit except that the plug outlet for a gas range may also be supplied by this circuit.

See section on services for more detail.

Washroom - Circuit for washer outlet may also supply the receptacle at the wash basin because it is in the laundry room. GFCI protection is required for the receptacle at the wash basin

Carport or Garage - At least one duplex receptacle is required for each car space. No other loads permitted on this circuit EXCEPT garage or carport lights and garage door opener.

Carport or garage lighting may be supplied by any nearby lighting circuit or by the carport or garage plug circuit, but the carport or garage plug circuit may supply only the carport or garage lights, plugs, and door opener, but not any other lights or plugs.

The Code Appendix B and its Definitions section (Location) identify the carport plug outlet location as an indoor location when applying Rules 26-710(n) and 26-714(b). This means that the carport plug outlet need not be GFCI protected. This is the strictly technical interpretation. The fact is that a carport plug outlet is freely accessible for use with electrical appliances normally used outdoors and should (not must) for safety reasons be GFCI protected.

Any wall space 900 mm (35.5 in.) or greater in width must have a plug outlet.

Note - This short space requires a plug outlet. The longest wall length is only 600 mm (24 in.) but measured into the corner, (along the floor line) as required by Rule 26-712(a) the length is 1140 mm (45 in.). This is more than 1 m therefore a plug outlet is required - it's the law.

Vacuum System

Double carport

Storage room

Family Room

Washer

Dryer

Micro wave oven

Table

Kitchen

Range

Fridge

dn

Entry

Dining room

Living room

Outdoor receptacles
- Separate circuit is required.
- Max. 12 outlets per circuit.
- Duplex type required .
- GFCI type circuit breaker or receptacle must be used.

Notes

- Only plug outlets and a few light outlets are shown

- Bedrooms are not shown - plug outlets in these rooms are spaced as in a living room or family room.

- Basement wiring is not shown.

BATHROOM PLUG OUTLET

Rule 26-710(f)(g) requires at least one plug outlet in each bathroom. To comply with the rules this outlet must:

•   Be equipped with at least one duplex plug outlet which must be located within 1 m (39.4 in.) of one of the wash basins in the bathroom, and

•   Be located at least 1 m (39.4 in.) away from the bathtub or shower stall. This is a horizontal distance between the outlet and a shower stall or bathtub as illustrated below; and

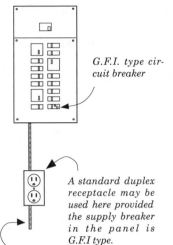

G.F.I. type circuit breaker

A standard duplex receptacle may be used here provided the supply breaker in the panel is G.F.I type.

This G.F.I. protected circuit may continue on to supply 11 more outlets

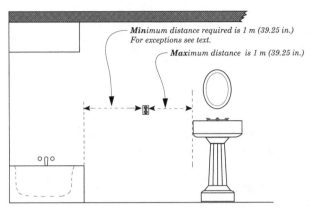

Minimum distance required is 1 m (39.25 in.). For exceptions see text.

Maximum distance is 1 m (39.25 in.)

Rule 26-710(g) - Where it is not possible (the subrule says "practicable") to locate the bathroom plug at least 1 m from the tub or shower it can be at a lesser distance, but it may not be closer than 50 cm (19.7 in.). This shorter distance is allowed only in the very small bathrooms where the 1 m distance is simply not possible.

• Be GFCI protected - Rule 26-700(11) This subrule requires GFCI protection for plug outlets in all bathrooms and washrooms. It may be either a GFCI type plug receptacle; or it may be an ordinary duplex plug receptacle that is supplied from a special type circuit breaker (GFCI) in the service panel. This circuit breaker is called a Class A Ground Fault Circuit Interrupter. These breakers mount in the service panel as ordinary breakers. If you are using an older type circuit breaker panel make sure they are available for the particular service panel you are using.

This subrule is very broad in its scope when applied according to the definition for bathrooms and washrooms. Those definitions do not really refer to a bathroom or to a washroom, but to rooms that contain a tub, shower stall, or washbasin. The room contains this appliance. It could be any room in the house. For example, a large bedroom with a tub is included in this definition. The fact that there is also a bed in the same room is irrelevant. This means that bathroom rules apply wherever any appliances, such as a tub or shower stall or a wash basin are being installed. Washroom rules apply wherever there is a washbasin; any room will do. Any plug outlets within 1.5 m (59 in.) of a bathtub, shower stall or washbasin must be GFCI protected, Rule 26-700(11), and

• Be Supplied with any nearby lighting circuit which has 11 or fewer outlets. Note - Bathroom / Laundry Room Combination - Rule 26-700(11) - This is an important detail. See "Laundry in Bathroom," p. 91.

WASHROOM PLUG OUTLET

The definition, in Section 0 of your Code book, says a washroom means "a room that contains a wash basin(s) and that may contain a water closet(s) but without bathing or showering facilities." As noted above a wash basin in a bedroom makes the bedroom a washroom and nobody is arguing about the need for a wash basin in a bedroom, but there are rules to be observed, I mean electrical rules. You may also want to do your shaving there too, in the morning, and you will need a power outlet. Washrooms (any room) equipped with a wash basin(s) must:

• Rule 26-710(f) - Be equipped with at least one duplex plug outlet which must be located within 1 m (39.4 in.) of one of the wash basins in that room.

• The rules do not require a separate circuit. This outlet may be connected to any nearby lighting circuit, that does not already supply 12 outlets.

• Have GFCI Protection - This outlet must be GFCI protected. Rule 26-700(11) requires this protection for all plug outlets within 1.5 m (59 in.) of a washbasin no matter where the washbasin is located. You may use a GFCI circuit breaker in the panel or you may use a GFCI type receptacle. In either case the rule permits other plug outlets and lights in other washrooms and bathrooms, bedrooms, hallways etc. to be protected with the same GFCI protective device. The maximum number of outlets must not exceed 12. See also above under bathroom plug outlets.

Note - If you plan to supply bedroom plug outlets with the same GFCI protected circuit you are using for the washroom or bathroom lights and plug outlets, please remember that the bedroom plug outlets require AFCI protection. AFCI protection can only be provided with a circuit breaker in the panel, see "Bedroom Plug Outlet," p. 77.

OTHER ROOMS OR AREAS - PLUG OUTLETS

STORAGE ROOMS

Rule 26-712(a) - Every finished room, including storage rooms, must have minimum wiring so that no point along the wall is more than 1.8 m (71 in.) from a plug receptacle.

Storage spaces such as areas under the stairways, attics or crawl spaces do not require a plug outlet. In fact it is safer without one, so that appliances must be unplugged before being stored.

AFCI protection is required for receptacles, but remember, you are not permitted to put receptacles in a closet.

CLOSETS, CUPBOARDS, CABINETS ETC.

Rule 26-710(i) prohibits plug outlets in these enclosures except for special cavities built for specific heating and non-heating type appliances. See "Appliance Storage Garage," p. 87 for details.

DINING ROOM

Rule 26-712(a) - Dining rooms which are not part of a kitchen must be wired as a living room so that no part along the wall is more than 1.8 m (71 in.) from a plug outlet. For dining or eating areas forming part of a kitchen, see "Kitchen Counter Outlets and Circuits," p. 85. AFCI protection is required for receptacles.

KITCHEN COUNTER PLUG OUTLETS

Kitchen plugs for counters including islands and peninsulas do not require AFCI.

Rule 26-700(11) The Rules permit three different options for the kitchen counter plug circuits. These are not given in any order of preference.

Option 1: 2 - 15 Amp breakers, 3-wire 15 amp cable and two split receptacles mounted along the back wall of the counter. This plug circuit arrangement is still approved for use on the kitchen counter just as they have been approved for the last number of years.

The circuit shown below can supply two split duplex receptacles. These are standard 15 amp duplex receptacles, except that they have a small break-away section of metal on the hot (brass colored) side terminal block as shown below. When this section of metal is broken off it disconnects the two halves of the duplex receptacle from each other so that each half can be connected to a different circuit. They share a common neutral (white) wire. The two circuit breakers supplying this three wire cable must be located side by side in the panel and be fitted with a tie bar (shown below) so that the operating handles are tied together and function as one. This is a very important safety feature. See p. 37 for details.

This small tab connects the two halves of the receptacle together. Normally this tab is left in place but for split receptacles on kitchen counters this gold coloured tab must be removed.

Remember, the Code permits two duplex receptacles on one 3-wire circuit. The upper half of each duplex receptacle is on one circuit and the lower half of each receptacle is on the other circuit in the 3-wire supply cable.

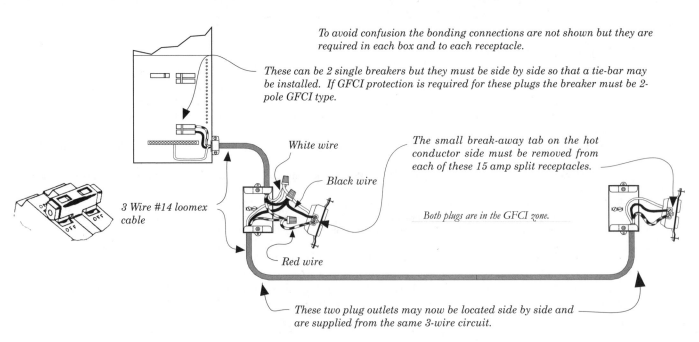

To avoid confusion the bonding connections are not shown but they are required in each box and to each receptacle.

These can be 2 single breakers but they must be side by side so that a tie-bar may be installed. If GFCI protection is required for these plugs the breaker must be 2-pole GFCI type.

White wire

Black wire

The small break-away tab on the hot conductor side must be removed from each of these 15 amp split receptacles.

3 Wire #14 loomex cable

Both plugs are in the GFCI zone.

Red wire

These two plug outlets may now be located side by side and are supplied from the same 3-wire circuit.

Caution - If one, or both, of the 15 amp split plug outlets illustrated above are located within the GFCI zone, i.e. within 1.5 m (59 in.) of the sink as illustrated on p. 85, then Rule 26-700(11) requires GFCI protection for these plugs. To provide GFCI protection for these outlets use a 2-pole 15 amp circuit breaker. This protection is not possible with two single GFCI type breakers nor is it possible with a split type GFCI receptacle.

Option 2: 1- 20 amp single pole breaker, 20 amp 2-wire cable and 2 - T-slot receptacles. Two different circuits are shown below. Where GFCI protection is required the upper circuit has a 20 amp GFCI type circuit breaker supplying two 2 - 20 amp T-slot receptacles. The lower circuit is supplied with a standard 20 amp circuit breaker and the GFCI protection, if it is required, is provided in the first outlet as shown.

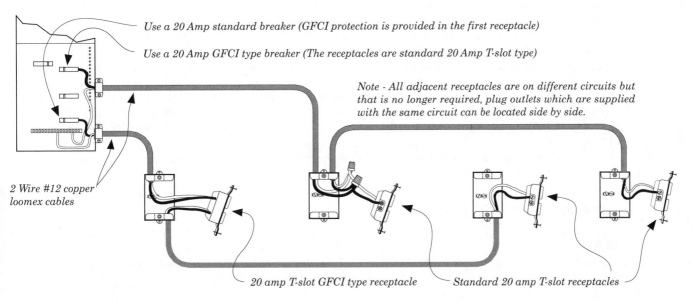

Use a 20 Amp standard breaker (GFCI protection is provided in the first receptacle)

Use a 20 Amp GFCI type breaker (The receptacles are standard 20 Amp T-slot type)

Note - All adjacent receptacles are on different circuits but that is no longer required, plug outlets which are supplied with the same circuit can be located side by side.

2 Wire #12 copper loomex cables

20 amp T-slot GFCI type receptacle   *Standard 20 amp T-slot receptacles*

If we choose to install these 20 amp receptacles on the kitchen counter there are a few things we must keep in mind:

- We must use 2-wire #12 copper supply cable.
- We must use a 20 amp fuse or breaker.
- We must use the type of 20 amp T-slot receptacle shown at left. Officially it is called a CSA Configuration 5-20RA receptacle.
- We can still supply two of these plugs with each 20 amp 2-wire circuit but if we do that we would actually reduce the total load capacity available at those two outlets. You see, with that arrangement there would be only one 20 amp breaker supplying two 20 amp, duplex outlets compared with two 15 amp breakers supplying two 15 amp duplex outlets when using the old option 1 circuit arrangement.
- The best way to gain an advantage with the 20 amp plug circuit is by installing a separate circuit for each outlet as shown below under option 3. With this arrangement you will have 20 A available at each outlet.

Option 3: 1 - 20 amp single pole breaker and 20 amp 2-wire copper cable is used to supply only one 20 amp T-slot receptacle. This arrangement is usable for any of the kitchen counter plugs, those which require GFCI protection and those that do not. Where GFCI protection is required for this circuit it can be with a single pole, 20 amp, circuit breaker in the panel, or with a 20 amp T slot GFCI type receptacle.

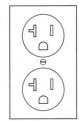

20 amp T-slot receptacle

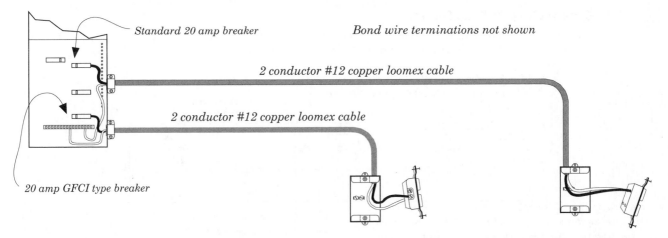

Standard 20 amp breaker   *Bond wire terminations not shown*

2 conductor #12 copper loomex cable

2 conductor #12 copper loomex cable

20 amp GFCI type breaker

Just one 20 amp T-slot standard type receptacle on a circuit.

Just one 20 amp T-slot GFCI type receptacle on a circuit.

Note the following details:

- Use single, 20 Amp circuit breakers.
- Use #12 copper cable for all wiring to 20 Amp receptacles.
- Use 20 Amp T-slot receptacles. Officially, they are called CSA Configuration 5-20RA receptacles.
- Use GFCI breakers or receptacles for outlets in the GFCI zone.
- Connect only two of these receptacles to a circuit.

These are the three different circuit arrangements possible. Some of these are by personal choice. For example, the location of GFCI protection either in the plug outlet or in the breaker in the panel way down there in the basement somewhere, is a personal choice. Other features in these three different circuit arrangements are mandated by Code. For example, the requirement for GFCI protection for plugs located within 1.5 m (59 in.) of the kitchen sink is a Code Rule. Other details, such as the two pole GFCI type circuit breaker, are required for technical reasons and some details are affected by the cost of the component parts. Choose your circuit arrangements carefully and try to imagine having to find that panel to reset the GFCI breaker at midnight just because you insisted on that last cup of tea.

20 amp Circuits for other outlets in the house? - Rule 26-710(b) makes it clear that all plug outlets, not lighting outlets, in the house could be wired with #12 copper cable and supplied with 20 amp breakers. But remember, this rule refers to plug outlets only, it does not mention lighting outlets. Rule 30-104(a) says that all lighting circuits in a dwelling unit must be protected by 15 amp fuses or breakers, not 20 amp. Most circuits in a residence supply a mixture of plugs and lighting to give better load diversity, and because it is more economical to do so. For this reason 20 amp circuits should (not must) only be used for special plug outlets.

Special circuits supplying only plug outlets, such as work shop outlets could be wired with #12 copper cable, a 20 amp T-slot receptacle, and supplied with a 20 amp breaker.

IMPORTANT POINTS TO REMEMBER

HEIGHT

The rules do not specify any height for these outlets except that they not be on the counter work surface facing up, Rule 26-710(c). They must be located on the wall above the normal splash level of the counter work surface.

MAXIMUM 2 RECEPTACLES PER CIRCUIT

Rule 26-722(b)(i) permits only two receptacles on a 15 amp and 20 amp kitchen counter plug circuits but there is an exception. See the Note regarding plug outlet locations for Disabled Persons, p. 88.

BEHIND AND IN FRONT OF THE KITCHEN SINK

Rule 26-712(e) - Plug receptacles must be located on the wall behind the counter and on either side of the kitchen sink but not behind it. It must not be necessary for appliance supply cords to pass over the sink. Just as important, plug outlets must not be located anywhere directly in front of the sink. Anyone in contact with a plug outlet while working in the sink could suddenly become highly charged and, damaged goods.

ADJACENT PLUG OUTLETS ON THE KITCHEN COUNTER

The Rule which required adjacent kitchen counter plugs to be supplied with different circuits has been deleted and good riddance it is. Adjacent kitchen counter plug outlets can now be on different circuits or on the same circuit as shown in the illustrations but remember, each counter plug circuit may supply only two receptacles except in the case of the additional outlets provided for use by persons with disabilities. This is covered on p. 88.

KITCHEN COUNTER PLUG OUTLETS

The cabinet layout shown below shows the normal circuit arrangements for kitchen counter plug outlets required by Code. Kitchen counter and fridge outlets do not require AFCI, but the clock outlet does.

These outlets are located within the GFCI zone and therefore must be GFCI protected as indicated on pages p. 107 and p. 82.

Clock outlet is no longer OK on a fridge circuit.

This plug outlet for the kitchen eating area must be on a separate circuit used for no other load except that it may also supply the plug outlet for a gas range. Counter plug circuits may supply only kitchen counter plug outlets. They may not supply any other outlets.

Use 15 amp breaker, #14 cable, and standard receptacle, or use 20 amp breaker, #12 cable, and 20 amp T-slot receptacles. Before installing a 20 amp circuit, check if 20 amp T-slot clock receptacle is available. GFCI protection is not required for these outlets.

Yes, these are 20 amp receptacles protected with a 20 amp breaker or fuse, and yes, these receptacles will accept a standard parallel blade 15 amp attachment cap. It will not be necessary to change the attachment caps on the appliances we now use in our kitchens. We will have various small kitchen appliances, with very small cords and which were designed and approved for use on 15 amp circuits, but will now be operating on 20 amp circuits. The fact that this arrangement has been used successfully for many years in The United States should convince us that this is a safe and acceptable practice.

POLARIZATION

This is an important detail. You will notice the receptacle has a brass terminal screw and a chrome plated terminal screw. Be sure to connect the black, or sometimes the red wire, to the brass terminal screw and the white neutral conductor to the chrome plated terminal screw. The Inspector has a little tester they use to check this connection without removing any cover etc. If you have connected incorrectly, they will very likely find it.

KITCHEN PLUGS IN GENERAL

According to Rule 26-712, 26-722, and 26-724, kitchen plug outlets are required as follows:

PLAN VIEW OF A KITCHEN

Below is the floor plan of a kitchen designed to show detail on what is required in a modern kitchen. The location of each of the plugs is as required by the Code, it shows the minimum number of plugs and circuits required for the counter space in this kitchen. Obviously the microwave plug shown in the illustration is not required if there is no special cavity provided for that oven. If the microwave is located on the kitchen counter the special plug outlet is not required, the oven could be plugged into any nearby counter plug outlet. The same is true for the manufactured appliance garage. If you choose to not install a plug outlet in that garage then the special switched plug outlet shown, and the CSA certification, would also not be required. Generally, the circuitry shown represents what is required by the Rules. Use this as a template to plan the wiring in your kitchen.

The arrangement of outlets and circuits on the kitchen counter work surface is illustrated below.

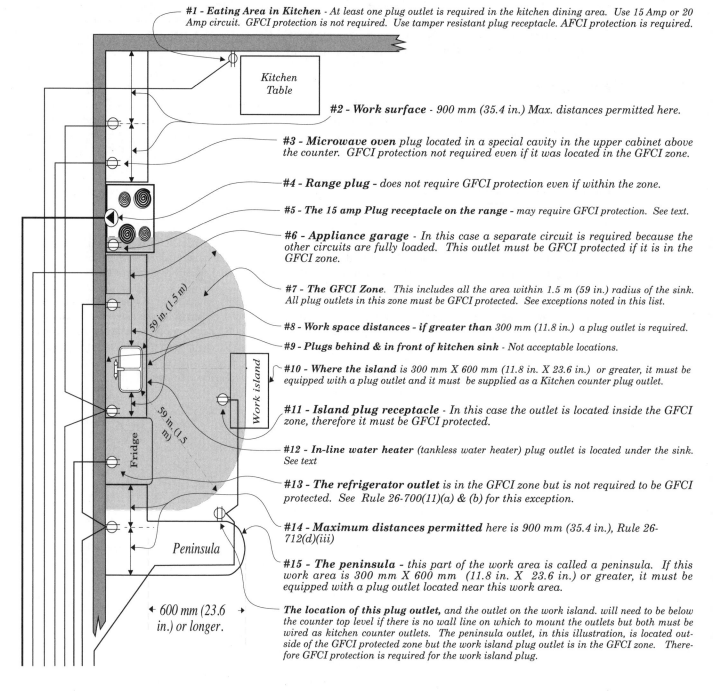

#1 - Eating Area in Kitchen - *At least one plug outlet is required in the kitchen dining area. Use 15 Amp or 20 Amp circuit. GFCI protection is not required. Use tamper resistant plug receptacle. AFCI protection is required.*

#2 - Work surface - *900 mm (35.4 in.) Max. distances permitted here.*

#3 - Microwave oven *plug located in a special cavity in the upper cabinet above the counter. GFCI protection not required even if it was located in the GFCI zone.*

#4 - Range plug - *does not require GFCI protection even if within the zone.*

#5 - The 15 amp Plug receptacle on the range - *may require GFCI protection. See text.*

#6 - Appliance garage - *In this case a separate circuit is required because the other circuits are fully loaded. This outlet must be GFCI protected if it is in the GFCI zone.*

#7 - The GFCI Zone. *This includes all the area within 1.5 m (59 in.) radius of the sink. All plug outlets in this zone must be GFCI protected. See exceptions noted in this list.*

#8 - Work space distances - if greater than *300 mm (11.8 in.) a plug outlet is required.*

#9 - Plugs behind & in front of kitchen sink - *Not acceptable locations.*

#10 - Where the island *is 300 mm X 600 mm (11.8 in. X 23.6 in.) or greater, it must be equipped with a plug outlet and it must be supplied as a Kitchen counter plug outlet.*

#11 - Island plug receptacle - *In this case the outlet is located inside the GFCI zone, therefore it must be GFCI protected.*

#12 - In-line water heater *(tankless water heater) plug outlet is located under the sink. See text*

#13 - The refrigerator outlet *is in the GFCI zone but is not required to be GFCI protected. See Rule 26-700(11)(a) & (b) for this exception.*

#14 - Maximum distances permitted *here is 900 mm (35.4 in.), Rule 26-712(d)(iii)*

#15 - The peninsula - *this part of the work area is called a peninsula. If this work area is 300 mm X 600 mm (11.8 in. X 23.6 in.) or greater, it must be equipped with a plug outlet located near this work area.*

The location of this plug outlet, *and the outlet on the work island. will need to be below the counter top level if there is no wall line on which to mount the outlets but both must be wired as kitchen counter outlets. The peninsula outlet, in this illustration, is located outside of the GFCI protected zone but the work island plug outlet is in the GFCI zone. Therefore GFCI protection is required for the work island plug.*

The following notes are arranged in the order they appear in the illustration above:

(1). Eating Area in Kitchen - If there is an eating area in the kitchen Rule 26-712(d)(6) requires a duplex plug outlet in that area and Rule 26-722(e) says this plug outlet must be supplied with a circuit used for no other purpose. Except that a plug outlet for the gas range may also be supplied with this circuit as shown on p. 89. AFCI protection for receptacles is required.

(2). Work Surface - Rule 26-712(d)(3)(4) & (5) - As shown above, the counter work space is often divided into several isolated sections. Each of these isolated sections must be considered separately and each must be provided with either a 15 amp split duplex receptacle or a 20 Amp T-slot receptacle if it is 30 cm (11.8 in.) or longer. This measurement is along the back wall of the counter space. The reason for this requirement is to make all counter work surfaces properly accessible to a plug receptacle without the supply cord having to cross over sinks, ranges, etc. Make sure you have a sufficient number of outlets along the counter before covering. Your inspector may not know what or where your counter work surfaces will be but they will check this detail in the final inspection. Plan it carefully and remember that any changes in the counter work spaces may affect the location of one or more plug outlets. It is difficult and costly to relocate, or add more outlets later, after the finish material is in place.

Work surface does not include the area of a sink, range, fridge or similar appliance. Understandably, the cook and the dishwasher, might want to challenge this narrow definition of a work area.

(3). Microwave - Rule 26-722(d) - A separate circuit is required for a built-in microwave. The plug outlet for this oven should be located in the same special oven cavity in the kitchen cabinet. Make sure there is adequate ventilation provided for this oven. See also below "Appliance Storage Garage," p. 87 and "Other Enclosures," p. 87.

Note - If it is not a built-in microwave, i.e. if it stands on the kitchen counter, this rule does not apply. In that case this oven would be regarded as any other movable appliance on the counter which can be served by any nearby plug outlet along the counter work surface.

(4). Range Plug Outlet - The 50 Amp receptacle for the range is not required to be GFCI protected even if it falls within the GFCI zone.

(5). 15 amp Plug Receptacle on the Range - Rule 26-710(j) - Plug outlets on the range or those located in cabinets or cupboards are not acceptable as alternatives for wall outlets but if these fall within 1.5 m (59 in.) of the sink they will need GFCI protection.

Caution - Rule 26-700(11) requires GFCI protection for 15 Amp and 20 Amp receptacles which are located within 1.5 m (59 in.) of a kitchen sink, this is the kitchen GFCI zone. This Rule makes an exception for a receptacle designated for a stationary appliance in that location but makes no exception for the 15 amp receptacle on the range. This receptacle can be used for any purpose in the same way any of the wall mounted receptacles are used and therefore must be GFCI protected if it falls within the GFCI zone. This requirement may be very difficult to comply with because the range may not be in place until some time after the final inspection of the house.

According to Rule 26-710(j) the receptacle on the range "shall not be considered as any of the receptacles required by this Rule." This disqualifier simply means that the range receptacle does not take the place of any requirement for wall mounted receptacles referenced in Rule 26-710. However, it does not apply to, or have any impact on, Rule 26-700(11).

This problem will likely be corrected with a provincial amendment. Check with your Provincial Code amendments. In the meantime that range outlet is a problem if it falls in the GFCI zone. Check with your local Inspector.

(6). Appliance Storage Garage - Rules 26-710(h) & (i) - Some modern kitchen cabinets provide cabinet space, (special cavities, for use as an appliance garage) inside the cabinets, or mounted on the kitchen counter, to store appliances normally left standing on the kitchen counter. If it is intended to install one or more plug receptacles in this garage, a dangerous situation could develop if the appliances can be stored while still plugged in. To prevent this possible dangerous situation the Rules require the following:

Subrule (i) permits a plug receptacle to be inside this appliance enclosure provided that "The receptacle is an integral part of a factory-built enclosure." This means that an appliance garage, complete with one or more plug outlets, must be a factory built assembly, not one built on site. The garage, complete with door operated switch and receptacles must be certified by CSA or by one of the other approval agencies, see p. 2.

Appliance garage must be factory assembled complete with door operated switch and one or more plug outlets.

This arrangement involves a door operated switch to control the plug receptacle located inside the enclosure. In order to ensure the correct and continuous dependable operation of the door operated switch, the subrule requires the garage assembly to be factory built and properly certified by CSA or one of the other certification agencies. An appliance garage is only a storage space for appliances, it does not refer to appliances which could be used while in such an enclosure.

Other enclosures - Subrules 26-710(h)(2)(3)(4)(5) refer to an installation such as an in-line water heater, or a compactor, a microwave, a cord connected range hood or fan, or a similar appliance certified for use when enclosed in a special cavity in the kitchen cabinet. The receptacle in the enclosure is "intended for use with specific type appliances, suitable for installation within the enclosure." Some of these appliances will each require a separate circuit but none will require a door operated switch because they are designed to be fully functional in their enclosure.

The receptacle intended for use with a microwave is usually in an open cavity above the kitchen counter. This cavity is rarely fitted with a door. Two things to watch out for: the microwave must be certified for use in a cavity. The concern is adequate ventilation for the oven. The other detail is the receptacle. It may be, in fact should be, located in that cavity. It does not need a switch to shut it off when not in use, but it does require a separate circuit, Rule 26-722(d).

(7). GFCI Protection Zone for Kitchen Plugs - The GFCI zone is the shaded area in the illustration, p. 85. It extends 59 in. (1.5 m) in all directions from the kitchen sink. With some exceptions noted here, all 15 amp and 20 amp plug outlets which fall within this zone, must be GFCI protected.

(8). Work Space Distances - Any counter space 30 cm (11.8 in.) long, or longer, must be equipped with a plug outlet. No point along the wall line above the work counter may be more than 90 cm (35.4 in.) from a plug receptacle. Any counter space which is 30 cm (11.8 in.) long, or longer, must be provided with one or more outlets. Any counter space less than 30 cm long is not required to have an outlet.

(9). Plug Outlets Behind, and in front of Kitchen Sink - Rule 26-712(e) - plug receptacles must be located on the wall behind the work surface on either side of the kitchen sink but not behind it. It must not be necessary for appliance supply cords to pass over the sink. Just as important, plug outlets must not be located anywhere directly in front of the sink. Anyone in contact with such a plug outlet while working in the sink could suddenly become damaged goods.

(10). Kitchen Island - Where the island is 30 cm X 60 cm (11.8 in. X 23.6 in.) or greater, the new Rule requires a receptacle on that island. Where the design of the island work area provides no suitable back splash board where a plug outlet can be mounted there is one solution, the receptacle can be mounted on a side or front surface of the cabinet directly below the work surface. Another solution is a raised type floor plug on the top surface of the work area but that may be regarded as an ungraceful impediment on the work surface. This plug outlet in this location must be tamper resistant type.

(11). Island Plug Receptacle - If this outlet is located within the GFCI zone it must be GFCI protected. If it is mounted on the side wall of the cabinet you will need a tamper resistant receptacle.

(12). In-line Water Heater (tankless water heater) - These are normally cord connected as permitted by Rule 26-744(9). The Rules do not require a plug outlet for this water heater unless you intend to install such a water heater. If it is intended to install an in-line water heater it will require a 15 amp single receptacle. Rule 26-710(i) will permit a plug outlet in the cabinet below the sink to supply this heater because it, the heater, is designed to be used in such an enclosure. Do not install a duplex receptacle in this location because that would make it possible to use the second outlet for another appliance and it could be stored while still connected to the power source. Such live appliance storage is considered a serious fire hazard.

A separate circuit should be installed for this heater. This is a water heater load and according to Rules 8-200(1)(a)(v); 8-302(2), it must be regarded as a continuous load. While there is no clear Code requirement for a separate circuit it is strongly recommended.

(13). Refrigerator Outlet - Rule 26-722(a) - A receptacle must be installed for each fridge. A 15 Amp non-split receptacle on a separate circuit can be used. This circuit may not be used to supply any other load except a clock outlet. GFCI protection is not required even if it falls within the zone because that plug outlet is provided for a stationary appliance designed for the location, Rule 26-700(11). Install a 15 amp duplex receptacle supplied with a single 15 A breaker and 2-wire loomex cable.

(14). Maximum Distance Permitted - is 90 cm (35.4 in.), Rule 26-712(d)(iii). No point along the wall line above the work counter may be more than 90 cm (approx. 36 in.) from a plug receptacle. This is the same as item #2 above.

(15). Peninsula - This is that long piece of kitchen counter that sticks out and impedes the smooth flow of traffic through the kitchen. That's probably not the only reason for that thing but since it is there Rule 26-712(d)(v) now requires at least one plug outlet be installed on it. When you build that kitchen don't leave any long projections sticking out anywhere because the Rule makers, when they find it, may insist that you put a plug on it. Seriously now, a peninsula type kitchen counter work area which is 60 cm (23.6 in.) or longer and its width is 30 cm (11.8 in.) or greater must now be equipped with a plug outlet. This outlet must be GFCI protected if it falls within the GFCI zone. In most cases the peninsula will provide no back-splash wall on which to mount a plug outlet. The one solution is to mount the outlet on the side of the cabinet, directly below the work surface. Another solution is a raised type floor plug on the top surface of the work area but that may be regarded as an ungraceful impediment on the work surface. This plug outlet in this location must be tamper resistant type.

OUTLET ACCESSIBILITY

Disabled Persons may have difficulty using counter plug outlets when they are located along the wall behind the counter. Rule 26-710(d) allows additional split plug outlets to be installed along the front or end wall of the lower cabinet to provide better access for the disabled. It should be noted that the rule does not require these outlets, it simply allows them in addition to the standard outlets. Like standard outlets, they do not require AFCI. The two subrules that deal with these special plugs are: Rule 26-722(d) permits these additional special plug outlets to be supplied from the normal kitchen counter plug circuits used to supply the normal kitchen counter plug outlets along the wall behind the counter, as illustrated below. This Rule allows us to connect two additional plugs onto each kitchen counter plug circuit. These special plug outlets, in fact, serve the same work spaces as the plug outlets above the counter so that it is not necessarily an increase in actual load on the normal circuits. The illustration below shows these special plug outlets. The outlet in the eating area must be AFCI protected.

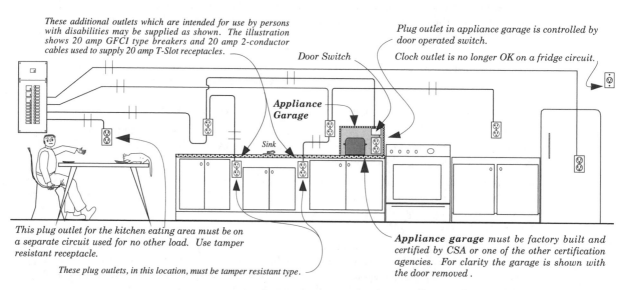

These additional outlets which are intended for use by persons with disabilities may be supplied as shown. The illustration shows 20 amp GFCI type breakers and 20 amp 2-conductor cables used to supply 20 amp T-Slot receptacles.

Door Switch

Appliance Garage

Sink

Plug outlet in appliance garage is controlled by door operated switch.

Clock outlet is no longer OK on a fridge circuit.

This plug outlet for the kitchen eating area must be on a separate circuit used for no other load. Use tamper resistant receptacle.

These plug outlets, in this location, must be tamper resistant type.

Appliance garage must be factory built and certified by CSA or one of the other certification agencies. For clarity the garage is shown with the door removed .

Rule 26-710(d) says that these special counter plug outlets are in addition to those shown in the illustration above, they are not a substitute for the plug outlets normally required along the counter.

SHORT KITCHEN WORK SPACE

The short counter shown below has limited workspace - only one split receptacle is required. Rules 26-722(c); 26-712(d)(3) & 26-722(b) recognize there will be cases where only one counter plug is required. Rule 26-700(11) requires GFCI protection for this outlet. Rule 26-722(b)&(c) permits a 20 amp circuit and one T-Slot receptacle, or one 2-pole, 15 amp GFCI type breaker and 3-wire circuit with one split receptacle for this location, as shown below. AFCI protection is required for receptacles in the kitchen table area and dining room.

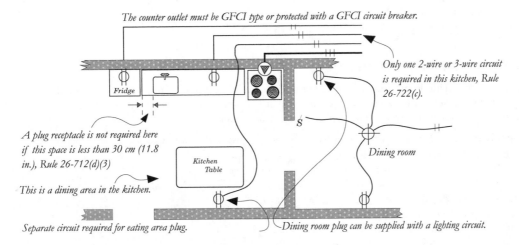

The counter outlet must be GFCI type or protected with a GFCI circuit breaker.

Only one 2-wire or 3-wire circuit is required in this kitchen, Rule 26-722(c).

A plug receptacle is not required here if this space is less than 30 cm (11.8 in.), Rule 26-712(d)(3)

This is a dining area in the kitchen.

Dining room

Separate circuit required for eating area plug.

Dining room plug can be supplied with a lighting circuit.

In the illustration below the counter space is longer.

There are two separate work spaces on this kitchen counter and Rule 26-712(d)(3) requires a plug outlet in each work space which is 30 cm (11.8 in.) or longer. Two outlets are required, one on each side of the sink and these must be GFCI protected. We can use two separate 20 amp single pole GFCI protected circuits as shown or we can use two separate 15 am 3-wire split receptacle circuits each supplied with a 2-pole 15 amp GFCI type breaker.

Where the counter is longer there will be more usable work spaces and therefore more receptacles are required. In the illustration below we had to add one more receptacle for the additional work space. Please note, that even though we are normally permitted to connect two 20 amp T-slot receptacles to a 2-wire 20 amp circuit or two 15 amp split receptacles on a 3-wire circuit, we could not connect these two receptacles to the same circuit in this case because Rule 26-722(b) requires at least two circuits whenever two or more receptacles are required. Because both of these plugs are in the GFCI zone, GFCI protection must be provided for both of these plugs.

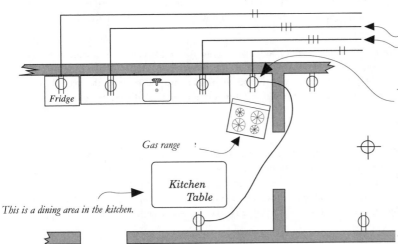

In this case each counter plug must be supplied with its own 3-wire 15 amp split receptacle, or 20 amp 2-wire T-slot receptacle. GFCI protection is required for these outlets, Rule 26-722(b).

Gas range - If gas supply piping or a gas connection outlet is provided for a free standing gas range, a plug outlet must be installed in the middle of the range space and within 13 cm (5.1 in.) of the floor. This is for control and illumination of the gas range. Locate this 15 amp plug for the gas range and the 50 amp plug for the electric range near the center of the wall space behind the range and not more than 13 cm (5.1 in.) to the center of the outlets above the floor. See also p. 115. This 15 amp range outlet may be supplied with the kitchen dining area plug circuit. These outlets must be flush mounted wherever practicable, Rules 26-712 and 26-744. If there is a built-in gas or electric cook top or oven installed, the rules do not require a plug outlet for a free standing electric range. If built-in gas or electric cooking facility is not provided, a plug outlet for a free standing electric range must be installed even if there is, or will be, piping for a gas range.

This is a dining area in the kitchen.

This eating area plug must be on its own circuit except that the gas range outlet may also be supplied with this circuit. Use a tamper resistant receptacle in this location.

GAS RANGE POWER SUPPLY

AFCI protection is required for the gas range power supply.

Rule 26-712(d) This subrule is illustrated above. It requires that where gas supply piping or a gas connection outlet is provided for a free standing gas range, we must also provide a standard 15 amp plug receptacle for the control voltage in that gas range.

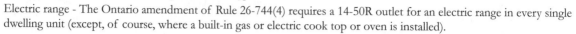

Rule 26-722(e) says this plug outlet may be supplied by the kitchen eating area plug circuit. The subrule does not say so but it is probably safe to assume this gas range plug could be supplied from any nearby lighting circuit as well. This plug outlet behind the range must be:

- Flush mounted on the wall behind the intended range location.
- Located near the midpoint measured horizontally along the back wall of the range cavity.
- Not more than 13 cm (5.1 in.) above the finished floor level.
- This receptacle need not be Tamper Resistant type.

Electric range - The Ontario amendment of Rule 26-744(4) requires a 14-50R outlet for an electric range in every single dwelling unit (except, of course, where a built-in gas or electric cook top or oven is installed).

For all provinces that have adopted the Code, a 14-50R outlet is required for free-standing ranges as well as where gas piping is provided for a free standing gas range. This is to ensure that an electric range can be safely connected and disconnected by each new owner or occupant. See also "Domestic Ranges," p. 115.

ADDITIONAL NOTES ON KITCHEN PLUGS

(1). Box size is important. Remember to use the correct column in the Table on p. 127 if you are installing #12 loomex cable. And if the receptacles you plan to install are more than 1 in. deep, larger boxes may be required. See p. 3. Make certain the boxes are deep enough for all the stuff you plan to put into them.

(2). A Split Receptacle - Means a duplex receptacle intended for use with 3-wire circuits.

A split duplex receptacle is a standard duplex receptacle, except that it has a small break-away section of metal on the hot (brass colored) side terminal block as shown. When this section of metal is broken off it disconnects the two halves of the duplex receptacle from each other so that each half can be connected to a different circuit. They share a common neutral (white) wire.

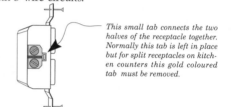

This small tab connects the two halves of the receptacle together. Normally this tab is left in place but for split receptacles on kitchen counters this gold coloured tab must be removed.

The Rules permit either these 15 Amp 3-wire, split duplex receptacles or the 20 Amp 2-wire T-slot receptacles described above to be installed along the wall behind the kitchen counter work surfaces. Remember, the Code permits two split duplex receptacles on one 3-wire circuit as shown above. The upper half of each duplex receptacle is on one circuit and the lower half of each receptacle is on the other circuit in the 3-wire supply cable. For additional information, "16.   Branch Circuit Wiring" on p. 52. Please note that the 15 amp split circuit arrangement is not acceptable if either one of the two receptacles is located in the GFCI zone. See illustration on p. 85 for clarification of the GFCI zone. Please note that GFCI protection must be provided for any kitchen counter plugs which are located in the GFCI zone.

(3). Wet Bar - The electrical load in a wet bar cannot be compared with the electrical load in a kitchen. Therefore, the rules that refer to kitchen loads cannot be correctly applied to the loads in a wet bar. There is an exception, a refrigerator receptacle located anywhere in a residence, for example a wet bar, must be supplied with a separate circuit used for no other load except a recessed clock outlet, Rule 26-722(a) Other specific use loads such as for example, cooking appliances, must be appropriately supplied according to the actual load. Other general purpose outlets in a wet bar in a residence would be regarded the same as the wiring requirements in a living room.

(4). The 23rd Code edition states that MOST receptacles in a residence require AFCI protection. The exceptions are: kitchen counter, island and peninsula receptacles, the refrigerator receptacle, and the receptacle for a sump pump (if it is a single receptacle and is labelled as being for the sump pump). Also, if the circuit is protected by an outlet-type AFCI installed at the first outlet on the circuit, the branch circuit feeder must be installed in a metal-armoured type cable or a metallic conduit.

DINING AREAS

Kitchen Dining Area - Rule 26-712(d) - If there is an eating area in the kitchen Rule 26-712(d)(vi) requires a duplex plug outlet in that area and Rule 26-722(e) says this outlet must be supplied with a separate circuit used for no other purpose, but there is an exception. We are allowed to also supply the plug outlet for the gas range with this circuit if we locate this additional outlet as shown in the illustration, p. 90. A 2-wire #14 copper loomex cable supplying a 15 amp duplex receptacle is acceptable, or a 2-wire #12 copper cable supplying a 20 amp T-slot receptacle is also acceptable. Receptacles need AFCI protection.

DINING ROOM

Rule 26-712(a) - A dining room which is a separate room and not part of the kitchen must be wired similar to the living room, i.e. no point along the floor line of the walls may be more than 1.8 m (71 in.) from a plug outlet. These outlets may be supplied with any lighting circuit. AFCI protection is required for receptacles.

LAUNDRY ROOM OR AREA

Rules 26-710(e)(i) and (ii); 26-722 - At least one appliance plug outlet must be installed in the laundry room or area. This laundry room plug outlet:

- May be a standard 15 amp receptacle on a 2-wire 15 amp circuit, or it may be a 20 amp circuit using a 20 amp breaker, #12 copper cable and a 20 amp T-slot receptacle.
- Must be duplex type - a single receptacle is not acceptable in the laundry room. It may be, but does not need to be, a split receptacle.
- Use tamper resistant receptacle.
- May be at any convenient and accessible height in the laundry room.
- Must be supplied by a circuit used for no other purpose than to supply the one or more duplex appliance plug receptacles in the laundry room or area.
- Must have AFCI protection.

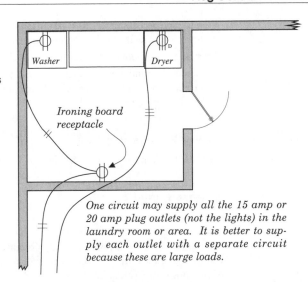

One circuit may supply all the 15 amp or 20 amp plug outlets (not the lights) in the laundry room or area. It is better to supply each outlet with a separate circuit because these are large loads.

The rule says we must provide at least one branch circuit solely for receptacles in the laundry room. According to this rule, two or more receptacles in the laundry area could be supplied by one circuit. It is better to install more circuits - one for each outlet. If there is sufficient space to do the ironing in this room you should, the rule does not say must, install a separate circuit for the iron.

UTILITY ROOM OR AREA

Rules 26-710(c), 26-722(c) - These rules require at least one duplex plug outlet in "each utility room." The term "utility room" is not defined in the Electrical Code but it is likely referring to the little room for the furnace. It is not referring to the laundry room because that room or area is dealt with separately under another subrule. It is not referring to the porch, or mud room, because those are covered under Rule 26-712(a)&(b), see below for details on porch wiring. It could be the furnace and water heater room but then what on earth would we do with a plug outlet in that room. It would serve no obvious purpose there. We have had fun with this requirement because so often this room is only a few in. larger than the furnace. In some cases it seems the furnace was installed then decorative walls were built around it and if one was quick enough the HWT could get in there too before the walls were built. Today we do not need a room to do the shoveling and the large coal bin, it's gone too. Still the Rule stands tall and demanding. It requires a plug receptacle, and it must be supplied with a circuit used for no other purpose, and yes, the receptacle must be tamper resistant type. AFCI protection is required for receptacles.

I have been poking fun at this Utility room requirement for many years and I do hope no one has taken offense. It seemed the one room that was just looking for calculated misunderstanding without offending anyone.

Since there is no definition of the term "utility room" then obviously no one can be sure when they are in that room. Such rooms are difficult to wire properly, but then, it would be just as difficult to prove it was not wired properly.

If the Inspector finds you with a "utility room" --which has not been properly wired, you could be in serious trouble, I suppose. See also "Freezer Outlet" on p. 92.

LAUNDRY IN BATHROOM

Rule 26-710(f) requires a plug outlet located within 1 m (39.4 in.) of the wash basin. Rule 26-700(11) requires this outlet to be GFCI protected, and Rule 26-710(g) requires this outlet to be located at least 1 m (39.4 in.) from the bathtub or shower stall where this is possible. Where this separation is not possible it may be reduced to not less than 50 cm (19.7 in.).

The receptacle for the washing machine must be located behind the appliance where it remains inaccessible for use with any other electrical appliances which may be used in that room. Rule 26-722b) requires a separate circuit for the washing machine. GFCI protection is not required. AFCI is required.

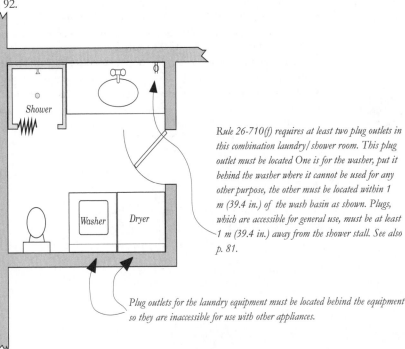

Rule 26-710(f) requires at least two plug outlets in this combination laundry/shower room. This plug outlet must be located One is for the washer, put it behind the washer where it cannot be used for any other purpose, the other must be located within 1 m (39.4 in.) of the wash basin as shown. Plugs, which are accessible for general use, must be at least 1 m (39.4 in.) away from the shower stall. See also p. 81.

Plug outlets for the laundry equipment must be located behind the equipment so they are inaccessible for use with other appliances.

Plug Outlets in Laundry Room/Bathroom Combination - Rules 26-710(f)(g) & (h) Laundry equipment may be installed in a bathroom provided the plug outlets for that equipment are properly located.

Provide a Plug Outlet at the wash basin as shown below. Provide Plug outlets for the washing machine.

If there is an ironing plug, locate it at least 39.4 in. (1 m) from the tub or shower stall. This receptacle must be GFCI type or be protected with a GFCI type circuit breaker in the panel and it should, (not must) be on its own circuit. Use only tamper resistant receptacles.

DRYER RECEPTACLES

See "Heavy Appliances" p. 114. AFCI protection is not required.

FREEZER OUTLET

The rules do not demand a separate circuit for the freezer but it is a good idea. We are still allowed to think of this one ourselves. The possible loss of a freezer full of meat because someone tripped the circuit breaker and forgot to reset it, makes this a good investment. Rule 26-722(c) requires a separate circuit for a plug outlet in the utility room. It may be this utility room outlet is intended for a freezer. This receptacle need not be tamper resistant if it is located behind the freezer, Rule 26-712(h). AFCI protection is required.

BALCONY OUTLET

Rule 26-712(b) Yes, the Code requires at least one a plug outlet on each balcony which is classified as unfinished Rule 26-710(a)'enclosed'. The Building Code refers to balconies with 'guards around' and others which are 'enclosed'. Enclosed balconies may well become an added living space with activity requiring electric power. The outlet may be supplied with any lighting circuit - maximum 12 outlets per circuit. See also "Unfinished Basement," p. 78. If the walls are finished then the balcony must have plug outlets similar to that required in a living room. An unfinished balcony requires only one plug outlet but a finished balcony must have plug outlets as in a living room. Use only tamper resistant receptacles. AFCI protection is required.

PORCH

Rule 26-712(b) - The lowly porch, it was discovered a few years ago and with its discovery came the requirement for a plug outlet. At least one plug outlet is required if the walls are not finished only if the thing is closed in. If the walls are finished, than the rules require plug outlets as in a living room. See also "Unfinished Basement," p. 78. Use only tamper resistant receptacles. The Rule does not require the walls to be finished, just enclosed. This outlet may be supplied with any lighting circuit - maximum 12 outlets per circuit. AFCI protection is required.

OUTDOOR PLUG OUTLETS

Rules 26-714(a); 26-714(b) & 26-724(a) - Basic requirement - Rule 26-702(4)26-714(a) - The rules require at least one duplex plug outlet which is readily accessible from ground or grade level for the use of appliances which need to be used outdoors. These outlets must:

- Be duplex type - single outlet is not acceptable, Rule 26-714(a).
- Be readily accessible, which means it must not be necessary to use chairs or ladders to reach the plug outlet, Rule 26-714(a). See definition in Section 0 of the Code.
- Tamper resistant type receptacles must be used for all outdoor plugs.
- Be supplied with a circuit used solely for this one or more outdoor plug outlets, Rule 26-724(a).
- Be a GFCI type receptacle or be supplied with a Class A ground fault circuit interrupter in the panel, Rule 26-710(n).
- In BC, be switched - This is recommended in Appendix B to Rule 26-714(a) "that outdoor receptacles be placed at both the front and rear of the house and that these be controlled by a switch located inside the house." Please note this is a recommendation by CSA it is not a requirement.

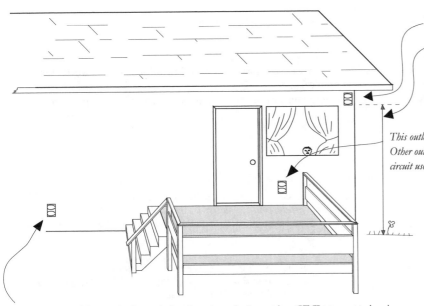

This outlet is for decorative lighting.

This outlet may be any height. If it is more than 2.5 m (98.5 in.) above grade it may be supplied by any lighting circuit inside the building. If it is less than 2.5 m (98.5 in.) above grade the outlet must be GFCI type or be supplied with a GFCI type circuit breaker. This outlet may be supplied by the same circuit used for other outdoor plug outlets.

This outlet must be GFCI type or supplied with a GFCI type circuit breaker. Other outdoor plug outlets may also be supplied with the same GFCI protected circuit used to supply this outlet.

This garden plug outlet must be GFCI type or be supplied with a GFCI type circuit breaker. Other outdoor plug outlets for special garden lighting may also be supplied with the same GFCI protected circuit used to supply this outlet.

See Garden Outlets, p. 114.

This is the preferred location for the required outdoor plug outlet. It must be a GFCI type receptacle or be supplied with a GFCI type circuit breaker. Other outdoor plug outlets may also be supplied with this GFCI protected circuit.

Use rigid metal or PVC conduit to protect the cable at this point.

NOTES AND BEWARES

(1). Sun deck Plug Outlets - Rules 26-710(o), 26-714(a) & 26-724(a) - If the deck is low and the plug outlets are within 98.5 in. (2.5 m) of ground or grade level the plug outlets must be GFCI protected with either a circuit breaker in the panel or a GFCI type receptacle. These are outdoor plug outlets and they must be supplied with one or more circuits which supply only outdoor plug outlets.

(2). Higher Sundecks - Rules 26-710(n), 26-714(a) & 26-726 - Where the outdoor outlet on the sun deck is more than 98.5 in. (2.5 m) above ground or grade level the rules regard it as not readily accessible and therefore is not affected by any of these three rules which refer to outdoor plug outlets. (Rule 26-702 still applies to these outlets where they are exposed to the weather, see p. 113 for details.) These outlets may still be supplied with the same GFCI circuit breaker as noted above, but they are not required to be on that circuit. They could be supplied with any nearby lighting circuit.

(3). Other Outdoor Plugs - Rule 26-710(n) & 26-724(a) - Refers to plug outlets located outdoors. This rule refers to "all receptacles" installed outdoors and which are within 2.5 m (98.5 in.) of grade level. Where these two conditions exist the receptacle must be GFCI protected. GFCI protection may be either in the circuit breaker or in the receptacle but, to avoid nuisance tripping where these receptacles are exposed to the weather, it may be better to provide GFCI protection in the circuit breaker in the panel. In , use tamper resistant receptacles.

(4). Decorative Lighting Outlets (Christmas Lighting) - These plug outlets need not be GFCI protected if they are at least 98.5 in. (2.5 m) above ground or grade level. If they are 98.5 in. or less above grade level they must be wired as described for any other outdoor plug outlet, Rule 26-724(a).

(5). Two Wire Cable - Use only 2-wire cable for this circuit. The GFCI circuit breaker will not work if wired with 3-wire cable where the third wire functions as a neutral conductor.

(6). Maximum Circuit load - Rule 26-724(a) requires at least one circuit for outdoor plug outlets. Where there are two or more outdoor plug outlets these may all be supplied with one circuit provided the total number does not exceed 12 outlets.

(7). AFCI protection is required for all branch circuits supplying 125 V receptacles (outlets) rated 20 A or less, except for circuits supplying bathrooms and washrooms and many kitchens circuits, all of which have their own requirements.

(8). GFCI Protection is required for the following plug outlets:
   - Rule 26-700(11) - All plug outlets - within 59 in. (1.5 m) of a wash basin, wherever it is located. See exceptions for the washing machine and dryer plugs in a combined washroom and laundry room, p. 91.
   - All carport plugs - See "Carport Plug Outlets," p. 94.
   - All outdoor plugs - which are on the outside of a single family dwelling or garage and those plug outlets in the garden which are within 98.5 in. (2.5 m) of grade, Rule 26-714(b) and Rule 26-710(n).
   - Bathroom light switch - In cases where the 1 m (39.4 in.)distance between a switch and a bathtub or shower stall is not possible Rule 30-320(3)(b) will permit a shorter distance but nothing less than 50 cm (19.7 in.) and then only if that circuit is GFCI protected.

CARPORT-ONLY PLUG OUTLETS

Rules 26-714(b) & (c), 26-724(b) - At least one plug outlet must be installed for each car space in the carport.

This outlet must:

- Be duplex type - single outlet is not acceptable.
- Be installed, one in each car space. The rule does not say 'in' each car space it says 'for' each car space. However, for convenience, it should (it is not a requirement), be in each car space.
- Be supplied with a circuit used solely for these outlets located in a carport except that the carport lighting may also be supplied with this circuit.
- Be Tamper Resistant type receptacle.
- AFCI protection is required.

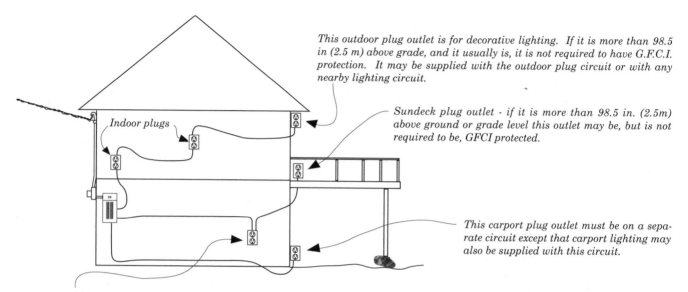

This outdoor plug outlet is for decorative lighting. If it is more than 98.5 in (2.5 m) above grade, and it usually is, it is not required to have G.F.C.I. protection. It may be supplied with the outdoor plug circuit or with any nearby lighting circuit.

Sundeck plug outlet - if it is more than 98.5 in. (2.5m) above ground or grade level this outlet may be, but is not required to be, GFCI protected.

Indoor plugs

This carport plug outlet must be on a separate circuit except that carport lighting may also be supplied with this circuit.

*This outdoor plug outlet must be accessible from grade and must be supplied with a separate circuit used for no other load except that there may be as many as 12 **outdoor plug outlets** supplied with this circuit.*

Use a GFCI type circuit breaker in the panel or use GFCI type receptacles.

Sundeck plug outlet may be supplied with this circuit but need not be if it is more than 98.5 in. above grade level.

NOTES

(1). Is GFCI Protection Required for the Carport Plug? - Yes it is. - well, sort of. Although this is a carport plug outlet it is in fact facing outdoors just as any other outdoor plug does, and could be used as any other outdoor plug outlet. Therefore, Rule 26-714(b) should be applied. No, the rules do not specifically require GFCI protection for this outlet but it is prudent to install it for all outdoor plug outlets. According to Rule 26-710(n) "all receptacles installed outdoors and within 2.5 m (98.5 in.) of finished grade" must be GFCI protected.

(2). Separate Circuit Required - Rule 26-724(b) says carport plug outlets must be supplied with a circuit used for no other purpose except that the carport lights may also be supplied with this circuit.

GARAGE PLUG OUTLETS

Rules 26-714(c) & 26-724(b) - At least one appliance plug outlet must be installed in each car space in a garage.

This outlet must:

- Be duplex type - single receptacle is not acceptable.
- Be installed so that there is a plug outlet in each car space. To be truthful, the rule does not say "in" each car space, it says "for" each car space. However, the intent seems to be that each plug outlet should be located in its own car space.
- Be supplied with a circuit used solely for the plug outlets located in the garage except that garage light outlets and garage door openers may also be connected to this circuit.
- Be Tamper Resistant type receptacle.

GFCI protection - The garage plug outlets are not required to be GFCI protected. AFCI protection is definitely required for an attached garage. For a detached garage, there is debate because it is not part of or attached to the dwelling. Check with your local Inspector.

CIRCUIT REQUIREMENTS

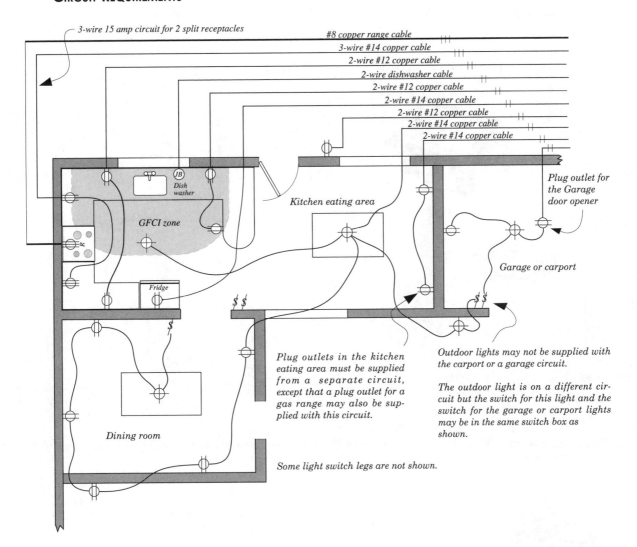

3-wire 15 amp circuit for 2 split receptacles

#8 copper range cable
3-wire #14 copper cable
2-wire #12 copper cable
2-wire dishwasher cable
2-wire #12 copper cable
2-wire #14 copper cable
2-wire #12 copper cable
2-wire #14 copper cable
2-wire #14 copper cable

Plug outlet for the Garage door opener

JB
Dish washer

Kitchen eating area

GFCI zone

Garage or carport

Fridge

Plug outlets in the kitchen eating area must be supplied from a separate circuit, except that a plug outlet for a gas range may also be supplied with this circuit.

Outdoor lights may not be supplied with the carport or a garage circuit.

The outdoor light is on a different circuit but the switch for this light and the switch for the garage or carport lights may be in the same switch box as shown.

Dining room

Some light switch legs are not shown.

DOORBELL TRANSFORMER

Rule 16-200, 16-204 - Type - Must be a CSA certified Class II transformer. This is usually die stamped somewhere on the transformer. This Class II label, is very important. It means the transformer is designed so that it will not be a fire hazard even if improperly wired.

Circuit - may be connected to any lighting circuit.

Location - Watch this one. This transformer must be located where it will remain accessible. This means it may not be inside a finished wall or ceiling where there is no access for maintenance or replacement later.

Caution - Do not mount this transformer inside the service panel. It may be nippled into the side of the branch circuit panel provided the wall finish is kept back to keep it exposed and accessible. If you locate it here it must then be supplied with a circuit breaker used for no other load. It is not correct to connect two wires to a breaker or to splice the transformer primary leads onto another circuit conductor in the panel, Rule 12-3032(1).

The furnace room or basement workshop area light outlet box is usually a good location for this transformer.

Cable - Type LVT cable may be used.

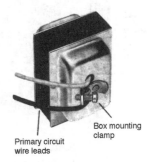

Box mounting clamp

Primary circuit wire leads

19. TYPES OF BOXES

Don't forget, some of these boxes must be in a Vapour Barrier enclosure. See p. 65 for details.

TYPE

There are many types of boxes available but only a few are in common use today.

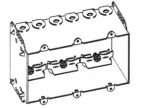

These boxes do not require the additional outer vapour barrier box as shown on p. 65.

Sectional metal box. This box can be dismantled.

Fixed Metal 3-gang switch box

Light outlet box

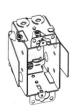

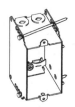

Knockout filler. Use these to close any open & unused knock-out holes.

Blank cover Make sure all junction boxes are covered.

Handy box for surface work

Switch or plug box for additional wiring after the wall finish is in place.

Rigid metal box

Extension ring

For bonding metal gang boxes see p. 101.

Internal Cable Clamp - Use the clamp properly. Where the cable enters the back or top knockout, as shown, it must emerge at the lower edge. It is not correct to double it back over the sharp edge, as shown.

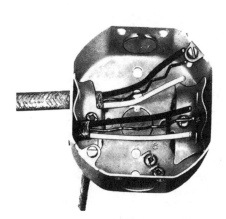

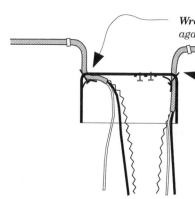

Wrong way! *Cable can be damaged by clamping it against the sharp edge of the knock out hole.*

Right way! *If the cable enters the back wall of the outlet box it must emerge along the side wall inside the box as shown.*

Saucer Boxes - The restrictions on the use of this box have been removed. This box may be used anywhere that similar deeper boxes are permitted. See box fill table, p. 105.

Do not use the center KO hole in this box unless the fixture you plan to connect to this outlet is the simple lamp-holder type shown.

Most light fixtures use a mounting strap to hold the fixture in place. There is usually a long hollow bolt which runs through the center of the fixture base and into the mounting strap. It's this long hollow bolt that may cause trouble if it extends too far into the shallow box because it is directly in line with the center KO hole in the box. If your supply cable enters the center KO hole it could be seriously damaged with this fixture mounting bolt. For this reason you should never enter a shallow box through the center KO hole. Always use one of the off center KO holes for cable entry. That center KO was not intended for cable entry but for special box mounting and heavy fixture support with a special box supporting bar.

These shallow boxes are often used at front and back doors when the outside wall finish is not smooth. As shown, the box is fastened directly to the outside rough sheathing. Because it is so shallow it need not be recessed into the sheathing. When the finish siding is installed the box may be shifted to match the boards so that it is fully recessed into the outer finish sheathing material. This eliminates the possibility of the outlet box being somewhere on the joint, between two boards, where it is difficult to fit the fixture properly and to seal (weatherproof) the opening around the fixture.

Box Loading - saucer boxes are very shallow, approximately 1/2 in. deep, and therefore may be used only at the end of a cable run. Only one 2-conductor #14 or #12 cable may enter this box. This means that you must run to the switch box first then to this light outlet box.

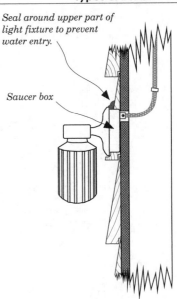

Seal around upper part of light fixture to prevent water entry.

Saucer box

BOX SUPPORT
Rules 12-3012, 12-3014

NAIL-ON TYPE

Most boxes, both plastic and metal, can be nailed onto a stud or joist and Rule 12-3012(5) says that supporting nails may even pass through the inside of a box provided they are hard against either the ends or back of the box so that they do not interfere with the conductors or connectors in the box. Nails must be driven in all the way - not just halfway, then bent over.

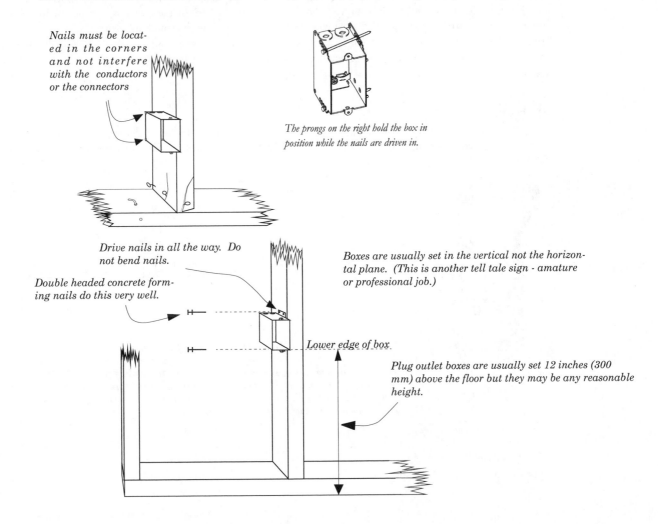

Nails must be located in the corners and not interfere with the conductors or the connectors

The prongs on the right hold the box in position while the nails are driven in.

Drive nails in all the way. Do not bend nails.

Double headed concrete forming nails do this very well.

Lower edge of box

Boxes are usually set in the vertical not the horizontal plane. (This is another tell tale sign - amature or professional job.)

Plug outlet boxes are usually set 12 inches (300 mm) above the floor but they may be any reasonable height.

SECTIONAL METAL GANG BOXES

Rule 12-3012(2) - Metal gang boxes may be supported with a brace, as shown below, or with wood backing.

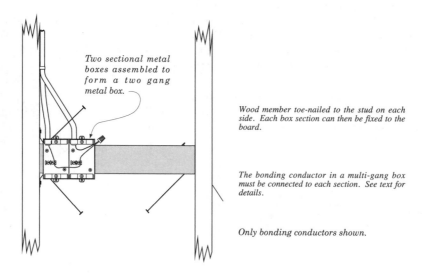

Two sectional metal boxes assembled to form a two gang metal box.

Wood member toe-nailed to the stud on each side. Each box section can then be fixed to the board.

The bonding conductor in a multi-gang box must be connected to each section. See text for details.

Only bonding conductors shown.

Metal sectional boxes can be easily transformed into multi-gang boxes by assembling sections until the desired box size is achieved. This is an advantage when a single sectional box has been installed and is already wired and a change in the wiring requires a two gang box. The conversion to a two gang box is easy but the additional support required for a sectional multi-gang box may be more work than replacing the single gang box with a fixed two gang box that can be fixed to a stud on only one side. The sectional multi-gang box is not stable enough for single side mounting, each section needs to be individually fixed to a wood member behind the box.

Metal sectional boxes must be secured with metal supports or have a header board fixed between the studs. Metal supports are the easiest to use. For metal sectional gang boxes use one of the metal side plates which were removed to form the gang box. Use the side which has the nail-on lugs and fasten one end to the back of the gang box with one of the bonding screws installed from the back of the box. The other end of the brace is nailed to the stud, as shown.

An alternative method is to fasten the gang box to a wood member installed behind the box. This is usually an unhappy experience because of the many screws protruding through the back of the box.

Caution - For bonding of sectional metal boxes, see "Bonding of Boxes," p. 99.

SET FLUSH

Rule 12-3018 - Set all boxes so that they are flush with the finished surface. Pay particular attention to the feature walls of wood paneling. When mounting boxes, don't forget to allow for the thickness of the wood strapping, in addition to the wall finish material.

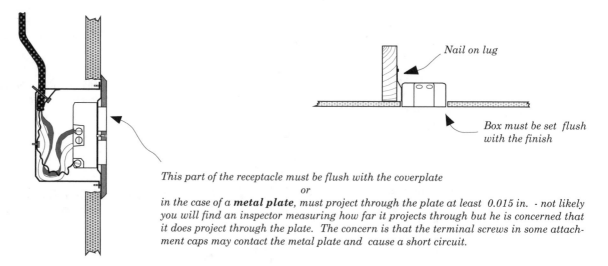

Nail on lug

Box must be set flush with the finish

This part of the receptacle must be flush with the coverplate
or
*in the case of a **metal plate**, must project through the plate at least 0.015 in. - not likely you will find an inspector measuring how far it projects through but he is concerned that it does project through the plate. The concern is that the terminal screws in some attachment caps may contact the metal plate and cause a short circuit.*

BOX EXTENDERS

Approved box extenders must be used where the outlet box is not flush with the finish.

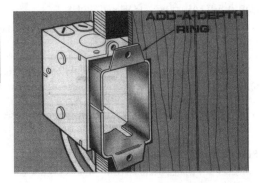

These are extension rings for use when the original box is set a long way back from the finished surface.

Where the box has been set for the wrong thickness of wall or ceiling finish and it is not flush as required, a simple box extension can be made with a section of plastic outlet box, as shown. Similar box extenders can now be purchased.

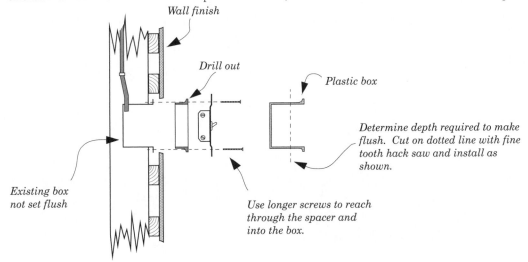

Wall finish

Drill out

Plastic box

Determine depth required to make flush. Cut on dotted line with fine tooth hack saw and install as shown.

Existing box not set flush

Use longer screws to reach through the spacer and into the box.

BONDING OF BOXES

Rules 12-526, 10-400, 10-808(2), 10-906 - The bare bonding conductor in the outlet box must be properly connected as shown below.

The bare wire should:

(1). First connect to each box as shown.
   • In sectional metal gang boxes (see p. 96) it connects to each section. Leave one long enough to loop around a bonding screw in each section then pigtail it to the all the others.
   • In fixed metal gang boxes, loop the Bond wire around one of the bonding screws and leave it long enough to pigtail it to all the other bond wires in that box.
   • In plastic gang boxes it connects to each metal strap inside the box unless these straps are already joined together by the manufacturer. In that case connect it to one bonding screw, then to all the other bond wires in that box.

(2). Next it connects to each bare wire entering the outlet box. See "Conductor Joints and Splices," p. 101 for comments on methods of joining these wires.

(3). Last, it connects to the green bonding screw on the base of the plug receptacle. This connection must be made with a pigtail, as shown.

PIGTAILS REQUIRED FOR BONDING CONNECTIONS

Rule 10-808(2) & 10-906(6) - As noted above, where there are two or more bare bond wires they must be properly joined and connected to both the box and the receptacle. The may not be twisted together and wrapped around a bonding screw. Rule 10-808(2) permits only one wire connected to any one bonding screw. If you have two or more bonding wires connect one to the bonding screw in the box but leave it long enough so that the other wires can be spliced to it as shown. It must be possible to disconnect a plug receptacle without interrupting or interfering with the bonding connections of downstream outlets on the circuit.

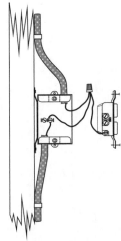

Pigtails Required for Neutral Wire Connections

Rule 4-034(4) - Neutral wire in 3-wire cable - Rule 4-034(4) requires a pigtail to make this connection in every case where 3-wire cable is used. Without the pigtail both neutral wires would need to be connected to the receptacle. With such a connection anyone removing the receptacle would automatically interrupt the balancing effect of the neutral so that other loads downstream would be subject to extreme fluctuation in voltage. Neutral conductor connections, when installing 3-wire cables, are even more important than the hot conductor connections. An opened, or failed, neutral connection can easily result in downstream outlets rated 115 volts, suddenly operating at more than 200 volts. Any electronic devices plugged into such an outlet is destroyed instantly. To prevent this situation the Code requires the connection to be made using pigtails as shown.

Pigtails not Required for Hot Conductor Connections

The rules do not require pigtails but where there are more than two hot conductors, for example, then the solution may be to use pigtails as shown in the illustrations.

CONDUCTOR JOINTS & SPLICES

Rules 4-034(4), 10-808(2), 12-506(1) - The illustration below shows a number of different kinds of outlets and the different connection methods required by code in each case.

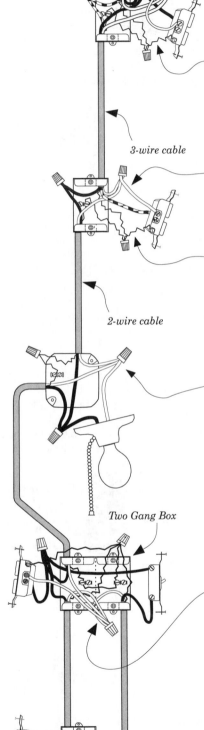

Home run to panel

3-wire cable

2-wire cable

Two Gang Box

Neutral wire in 3-wire cable - *A pigtail is required in every case. Without the pigtail both neutral wires would need to be connected to the receptacle. With such a connection anyone removing the receptacle would automatically interrupt the power to other loads downstream.*

Hot wires - *also require a pigtail, see BC Bulletin 26-1 - 0(1).*

Bare Bond wires - *Rules 10-808(2) & 10-906(6) requires a pigtail whenever a circuit continues on to supply other outlets downstream so that the receptacle can be removed **without disconnecting** downstream outlets from ground. Connect one bonding conductor to the box by taking the shortest route. Do not cut the wire, leave it long enough to make proper splices with other bond wires and a pigtail as shown.*

Neutral Wire - *This is still 3-wire cable and therefore, Rule 4-034(4) requires a pigtail.*

Hot Wires - *One connects directly to the receptacle the other is spliced black to black to supply other loads down stream.*

Bare Bond Wires - *Rules 10-808(2) & 10-906(6) require a pigtail because the circuit continues on to supply other outlets downstream.*

> *Sometimes an electrician will leave longer lengths of free conductor . He will complete the splice with a connector cap. Later, when he installs the receptacle he will simply loop the bare wire around the bonding screw . This eliminates the need for a pigtail and it is just as acceptable.*

Neutral Wires - *Pigtails are required here because this is a pull-chain type light fixture which has provision for terminating only one neutral wire. Ordinary keyless light fixtures usually have provision for terminating two neutrals and two hots. Where such fixtures are used pigtails are not required unless you are using 3-wire cables or you have more than two neutrals or two hots. Using pigtails to make any of these connections may not always be required by Code but it is still the much better way because it eliminates a lot of strain on the fixture or receptacle termination points.*

Hot Wires - *See under neutral wire above.*

Bond Wires - *This light fixture does not require a connection to the bond wire but all the bond wires must be spliced together as shown .*

Neutral Wires - *A pigtail is required to join the three neutrals and make the connection to the receptacle.*

Hot Wires - *A pigtail is required to connect the two devices and to splice the two hot conductors together. Make sure the hot conductors are connected to the brass coloured screws.*

Bare Bond Wires - *The bare bond wire in the supply cable should be left long enough to connect to the bonding screw in each box then continue on to a wire connector where all the other bare wires are spliced together. A pigtail is required here to connect the receptacle. Do not forget to connect the bond wire to each section of a multi-section metal gang box. One piece metal gang boxes require only one bond connection but multi-gang boxes consisting of a number of sections must have each section connected to the bond wire as shown.*

JOINTS & SPLICES IN BOXES

Rule 12-506(1) - Joints may be made only in outlet boxes or junction boxes. Junction boxes should be used very sparingly because you can get into more trouble with the Inspector when they find them. Junction boxes may not be buried in walls or in ceilings.

SOLDER OR MECHANICAL JOINTS & SPLICES

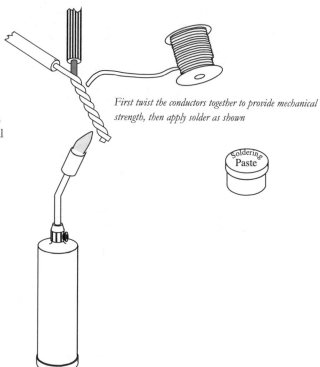

First twist the conductors together to provide mechanical strength, then apply solder as shown

Rules 10-808(2), 10-906(2), 12-112 - Solder - This is probably the best possible method of splicing circuit conductors.

It takes longer to make solder joints. Use a non-corrosive paste, usually 50/50 solder (50% tin 50% lead) for easy flowing and a minimum amount of heat. Then apply electrical tape. Build up a layer of tape equal to the insulation thickness of the conductor. This is needed, not for dielectric strength, but for mechanical protection.

Be sure to melt solder on the wire.

Caution - Do not solder bonding conductors - crimp-on type may be used. Taping is not required.

Don't try to cheat if you want to live to a ripe old age. If you are using the soldering method, use it, don't think you can get away with just twisting the wires together, then taping them without soldering the joint - your Inspector will find out.

Crimp-on - These are good if properly installed. Don't just gimble the connector with your pliers or side cutters and hope the Inspector will not see it. That is poor workmanship and is rejectable according to Rule 2-108. Use an approved crimping tool or use the twist-on type wire connectors.

Twist-on Insulator Caps - There are a number of different types and sizes of twist-on wire connectors available. You must use the correct size to make a good electrical connection. Check the marking on the carton to determine the number and size of wires permitted in each connector.

JUNCTION BOXES

Accessibility - Rules 12-3014(1), 12-112(3) - They must remain accessible. This means they may not be hidden inside a wall or ceiling or similar location.

Head Clearance - Rule 12-3014(2) - Where junction boxes are installed in an attic or crawl space, there must be at least 90 cm (35.4 in.) vertical space above this box to provide access for maintenance.

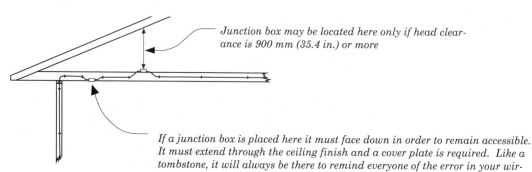

Junction box may be located here only if head clearance is 900 mm (35.4 in.) or more

If a junction box is placed here it must face down in order to remain accessible. It must extend through the ceiling finish and a cover plate is required. Like a tombstone, it will always be there to remind everyone of the error in your wiring job - not a good location.

Where to Use - Use junction boxes sparingly, only where you absolutely have to. Usually all your joints are made up in light, switch or plug outlet boxes.

WIRES IN BOX

RULE 12-3036

Free Conductor - At least 15 cm (6 in.), (it is better to leave 8 in.,) of free conductor must be left in the outlet box to allow joints to be made or fixtures to be connected in a workmanlike manner.

Cable Sheath - Rule 2-108 - Workmanship - The outer cable sheath should not project into the outlet box more than 0.5 in. past the connector. Remove this outer sheath as required before installing the cable in the connector. This sheath is very difficult to remove properly once the cable is actually installed.

Cable entries - Outlet boxes are equipped with four cable entries and each has a separate knock-out or pry-out and a cable clamp. Normally each cable entry is designed for just one cable. Do not attempt to install two cables in any one entry. When tempted to pull in a second cable through the same pry-out stop for a cup of strong coffee, Ontario Bulletin 12-19-9. Actually the Bulletin does not mention coffee, it just says, don't do it.

A cable connector designed for loomex cable should not be used for armoured cable because it cannot hold the armoured cable adequately.

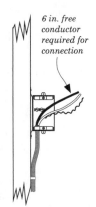

6 in. free conductor required for connection

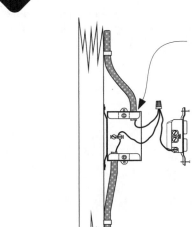

Cable entry holes are normally designed and intended for one cable only. These boxes have four cable entries, more than enough in most cases.

Do not over tighten these clamps, apply just enough pressure to hold the cable snugly (comfortably) in place.

BOX FILL

Rules 12-3034 and 12-3036 - Watch that box fill, it's tricky. This otherwise simple problem has been made difficult in the code book. We must count the number of wire connectors with insulated caps which we install in a box. We must subtract the space occupied by these connector caps from the space in the box. Of course, we are still required to note carefully the number of wires entering the box - deducting 1.5 cubic in. for each insulated #14 conductor. The bare conductor does not count. Then, the rules say the switch or receptacle in the box occupies space equal to two conductors. The tables below take all these factors into account.

The following should be carefully noted:

- Pigtails, (short lengths of wire used to connect things) do not count as box fill.
- Boxes may have internal or external cable clamps - it does not matter. The box fill is the same for both.
- Wires from light fixtures directly mounted and connected to the circuits in the box do not count as box fill.

NOTES ON BOX FILL TABLES

(1). Nominal Dimensions - Don't let the nominal box dimensions fool you. These are not the actual box sizes. It should be noted that some plastic (phenolic) box manufacturers keep the dimensions of their boxes close to the nominal. It is better to work from box volume than from its dimensions. Most box manufacturers now mark the cubic volume of their boxes with a die stamp inside the box. Look for these marked boxes.

(2). Various Combinations Given in the Table - In some combinations the Tables allow many more connector caps than could possibly be used for the number of wires in the box. It also shows other combinations where there are not enough connector caps for the number of wires in the box. Choose a combination which will permit you to install at least the number of wires you need and at least the number of connector caps you require. For example, if we are using #14 loomex cable and intend to run a 2-wire supply cable and two 2-wire load cables into a 2.⬛ in. deep switch box and plan to make 2 joints in this box the Table says NO! It's too full. We may install only 4-#14 wires and one connector in this box.

Only the insulated wires count as box fill; the bare bonding conductors do not count. In the above example the Table requires a deeper box. The Table shows a 2.5 in. deep box may contain the 5 - #14 wires and the two splices (connector caps) we need. We do not need to provide the extra space as far as the rules are concerned but this is the combination nearest to our needs and it gives us room for a little error in our planning.

(3). Insulated Cap - The rule says if we use these, we must reduce the number of wires in the box. The Tables starting on p. 105 take this into account.

This applies to insulated caps of all kinds.

(4). Tape Insulated Joints & Splices - The rule does not mention tape insulated joints or splices - perhaps it is because they occupy less space in the box. If you are using tape to insulate splices, you may add one conductor to the box fill indicated in the table. See "Conductor Joints and Splices," p. 101."

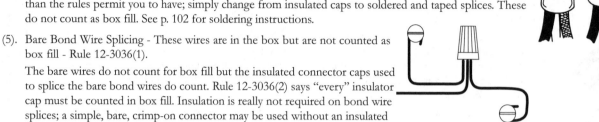

If you have used insulator caps and you find you have misjudged box size; you have more "things" in a box than the rules permit you to have; simply change from insulated caps to soldered and taped splices. These do not count as box fill. See p. 102 for soldering instructions.

(5). Bare Bond Wire Splicing - These wires are in the box but are not counted as box fill - Rule 12-3036(1).

The bare wires do not count for box fill but the insulated connector caps used to splice the bare bond wires do count. Rule 12-3036(2) says "every" insulator cap must be counted in box fill. Insulation is really not required on bond wire splices; a simple, bare, crimp-on connector may be used without an insulated cap and these do not count as box fill. "Box Fill" on p. 103.

(6). Remember that box fill is based on the physical size of an ordinary light switch and an ordinary duplex receptacle and there is nothing wrong with that provided we remember that certain dimmer switches and timer switches and certainly some of the GFCI type receptacles occupy more space. This is becoming more evident since manufacturers are required to include more safety features in the GFCI receptacles.

ACTUAL BOX FILL PERMITTED TABLE

Rule 12-3036

Box size and number of sections in the gang.	Volume of boxes in. (mL)	Using #14 cable this box may contain:	Using #12 cable this box may contain:
Single gang - metal box. Box size 3x2x1.5 in. deep	8 (131)	One switch or one plug outlets plus: Wires: 3, 3, 2, 2 Caps: 0, 0, 1, 2	One switch or plug outlets plus: Wires: 2, 2 Caps: 0, 1
2 gang - metal box. Box size 3x2x1.5 in. deep	16 (262)	Two switch or two plug outlets plus: Wires: 6, 6, 5, 4 Caps: 0, 1, 2, 3	Two switch or two plug outlets plus: Wires: 5, 5, 4 Caps: 0, 1, 2
3 gang - metal box. Box size 3x2x1.5 in. deep	24 (393)	Three switch or three plug outlets plus: Wires: 10, 10, 9, 8, 9 Caps: 0, 1, 2, 3, 4	Three switch or three plug outlets plus: Wires: 7, 7, 6, 5, 5 Caps: 0, 1, 2, 3, 4
Single gang - metal box. Box size 3x2x2 or 3x2x2.25 in. deep	10 (163)	One switch or one plug outlets plus: Wires: 4, 4, 3, 3, 2 Caps: 0, 1, 2, 3, 4	One switch or plug outlets plus: Wires: 3, 3, 2, 2 Caps: 0, 1, 2, 3
2 gang - metal box. Box size 3x2x2 or 3x2x2.25 in. deep	20 (327)	Two switch or two plug outlets plus: Wires: 9, 9, 8, 8, 7, 7 Caps: 0, 1, 2, 3, 4, 5	Two switch or two plug outlets plus: Wires: 7, 7, 6, 6, 5 Caps: 0, 1, 2, 3, 4
3 gang - metal box. Box size 3x2x2 or 3x2x2.25 in. deep	30 (491)	Three switch or three plug outlets plus: Wires: 14, 14, 13, 13, 12, 12 Caps: 0, 1, 2, 3, 4, 5	Three switch or three plug outlets plus: Wires: 11, 11, 10, 9, 9 Caps: 0, 1, 2, 3, 4, 5
Single gang - metal box. Box size 3x2x2.5 in. deep	12.5 (204)	One switch or one plug outlets plus: Wires: 6, 6, 5, 4, 4 Caps: 0, 1, 2, 3, 4	One switch or plug outlets plus: Wires: 5, 5, 4, 3, 3 Caps: 0, 1, 2, 3, 4
2 gang - metal box. Box size 3x2x2.5 in. deep	25 (409)	Two switch or two plug outlets plus: Wires: 12, 12, 11, 11, 10, 10 Caps: 0, 1, 2, 3, 4, 5	Two switch or two plug outlets plus: Wires: 10, 10, 9, 8, 8 Caps: 0, 1, 2, 3, 4, 5

Box size and number of sections in the gang.	Volume of boxes	Using #14 cable this box may contain:						Using #12 cable this box may contain					
3 gang - metal box. Box size 3x2x2.5 in. deep	37.5 (614)	One switch or one plug outlets plus: — Three switch or three plug outlets plus:						One switch or plug outlets plus: — Three switch or three plug outlets plus:					
		Wires	19	19	18	18	17	17	Wires	15	15	14	14	13	13
		Caps	0	1	2	3	4	5	Caps	0	1	2	3	4	5
Single gang - metal box. Box size 3x2x3 in. deep	15 (245)	One switch or one plug outlets plus:						One switch or plug outlets plus:					
		Wires	8	8	7	7	6	6	Wires	6	6	5	5	4	4
		Caps	0	1	2	3	4	5	Caps	0	1	2	3	4	5
2 gang - metal box. Box size 3x2x3 in. deep	30 (491)	Two switch or two plug outlets plus:						Two switch or two plug outlets plus:					
		Wires	16	16	15	15	14	14	Wires	13	13	12	12	11	11
		Caps	0	1	2	3	4	5	Caps	0	1	2	3	4	5
3 gang - metal box. Box size 3x2x2 in. deep	45 (737)	Three switch or three plug outlets plus:						Three switch or three plug outlets plus:					
		Wires	24	24	23	23	22	22	Wires	19	19	18	18	17	17
		Caps	0	1	2	3	4	5	Caps	0	1	2	3	4	5
Single gang - plastic box. These boxes usually have their volume clearly marked.	16 (262)	One switch or one plug outlets plus:						One switch or one plug outlets plus:					
		Wires	8	8	7	7	6	6	Wires	7	7	6	6	5	5
		Caps	0	1	2	3	4	5	Caps	0	1	2	3	4	5
Single gang - plastic box. These boxes usually have their volume clearly marked.	18 (294)	One switch or one plug outlets plus:						One switch or one plug outlets plus:					
		Wires	10	10	9	9	8	8	Wires	8	8	7	7	6	6
		Caps	0	1	2	3	4	5	Caps	0	1	2	3	4	5

GFCI RECEPTACLE DEPTH TABLE

1 1/8 in. deep (2.857 cm deep)

Volume of box	Using #14 cable					Using #12 cable				
	Wires / Caps	Wires / Caps	Wires / Caps	Wires / Caps	Wires / Caps	Wires / Caps	Wires / Caps	Wires / Caps	Wires / Caps	Wires / Caps
10 Cu. in (163 mL)	2 / 0	2 / 1				2 / 0	2 / 1			
12.5 Cu. in (204 mL)	4 / 0	4 / 1	3 / 2	3 / 3	2 / 4	3 / 0	3 / 1	2 / 2		2 / 4
15 Cu. in (245 mL)	6 / 0	6 / 1	5 / 2	5 / 3	4 / 4	5 / 0	5 / 1	4 / 2	4 / 3	3 / 4
16 Cu. in (262 mL)	6 / 0	6 / 1	5 / 2	5 / 3	4 / 4	5 / 0	5 / 1	4 / 2	4 / 3	3 / 4
18 Cu. in (294 mL)	8 / 0	8 / 1	7 / 2	7 / 3	6 / 4	7 / 0	7 / 1	6 / 2	6 / 3	5 / 4

1 1/4 in. deep (3.175 cm deep)

Volume of box	Using #14 cable					Using #12 cable				
	Wires / Caps	Wires / Caps	Wires / Caps	Wires / Caps	Wires / Caps	Wires / Caps	Wires / Caps	Wires / Caps	Wires / Caps	Wires / Caps
10 Cu. in (163 mL)	2 / 0	2 / 1				2 / 0	2 / 1			
12.5 Cu. in (204 mL)	4 / 0	4 / 1	3 / 2	3 / 3		3 / 0	3 / 1	2 / 2		2 / 4
15 Cu. in (245 mL)	5 / 0	5 / 1	4 / 2	4 / 3	3 / 4	5 / 0	5 / 1	4 / 2	4 / 3	3 / 4
16 Cu. in (262 mL)	7 / 0	7 / 1	6 / 2	6 / 3	5 / 4	5 / 0	5 / 1	4 / 2	4 / 3	3 / 4
18 Cu. in (294 mL)	7 / 0	7 / 1	6 / 2	6 / 3	5 / 4	7 / 0	7 / 1	6 / 2	6 / 3	5 / 4

GFCI receptacle — 1 3/8 in. deep (3.492 cm deep)

Box capacity	Wires	Caps	Wires	Caps	Wires	Caps	Wires	Caps	Wires	Caps
10 Cu. in (163 mL)	2	0	2	1	1	0	1	1		
12.5 Cu. in (204 mL)	3	0	3	1	2	2	3	0	3	1
15 Cu. in (245 mL)	5	0	5	1	4	2	4	3	3	4
16 Cu. in (262 mL)	6	0	6	1	5	2	4	3	4	4
18 Cu. in (294 mL)	7	0	7	1	6	2	6	3	5	4

GFCI receptacle — 1 1/2 in. deep (3.81 cm deep)

Box capacity	Wires	Caps	Wires	Caps	Wires	Caps	Wires	Caps	Wires	Caps
10 Cu. in (163 mL)	1	0	1	1						
12.5 Cu. in (204 mL)	3	0	3	1	2	2	3	0	3	1
15 Cu. in (245 mL)	5	0	5	1	4	2	4	3	3	4
16 Cu. in (262 mL)	5	0	5	1	4	2	4	3	3	4
18 Cu. in (294 mL)	7	0	7	1	6	2	6	3	5	4

LIGHT OUTLET BOXES, JUNCTION BOXES, OTHER BOXES TABLE

Box Dimensions	Volume of Box	Light outlet boxes Maximum combination wires and caps using #14 wire							Light outlet boxes Maximum combination wires and caps using #12 wire						
4x1-1/2 in. deep metal octagon box	15 (245 mL)	Wires	10	10	9	9	8	8		Wires	8	8	7	7	6	6
		Caps	0	1	2	3	4	5		Caps	0	1	2	3	4	5
4x2-1/2 in. deep metal octagon box	21 Cu. in. (344 mL)	Wires	14	14	13	13	12	12	11	Wires	12	12	11	11	10	10
		Caps	0	1	2	3	4	5	6	Caps	0	1	2	3	4	5
Shallow saucer type box Limited Use	5 (81 mL)	Wires	3	3	2					Wires	2	2				
		Caps	0	1	2					Caps	0	1				
4x2-1/2 in. deep metal octagon box Courtesy Nu-tek	26 (426 mL)	Wires	17	17	16	16	15	15	1	Wires	14	14	13	13	12	12
		Caps	0	1	2	3	4	5	6	Caps	0	1	2	3	4	5

Note - The above table is based on the fact that these boxes are used for light outlets and as junction boxes.
They will not contain any devices other than wires, connector caps and in some cases, cable connectors.

Below is Code Table 22. It also has the Imperial values as well as the metric values. Some of us still have a bit of difficulty with metric. We hear metric, quickly convert it to imperial, and then we understand. Is that illegal? I hope not because the whole idea motivating this book is to make the Electrical Code, the Safety Code, easy to understand.

Code Table 22

Size of Conductor AWG	Usable Space Required for Each Insulated Conductor Cubic Inches (Cubic Centimetres)
14	1.5 (24.6)
12	1.75 (28.7)
10	2.25 (36.9)
8	2.75 (45.1)
6	4.5 (73.7)

20. LIGHTING FIXTURES

SET FLUSH

Rule 12-3016 - Check first if the boxes worked out flush with the wall or ceiling finish. If not, see "Box Extenders," p. 98.

CONNECTIONS

Rules 30-108, 30-600 - When connecting light fixtures be sure to connect the neutral (the white or grey wire) to the screw shell of the lamp holder and the black or red (hot) wire to the center pin. The threaded metal portion of the lamp base must not ever be energized because of the danger of accidental contact with this portion by a person replacing a lamp. See p. 68 for details.

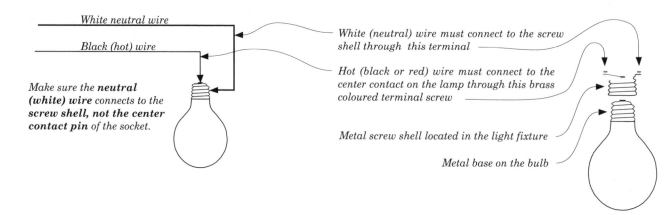

White neutral wire

Black (hot) wire

*Make sure the **neutral (white) wire** connects to the screw shell, not the center contact pin of the socket.*

White (neutral) wire must connect to the screw shell through this terminal

Hot (black or red) wire must connect to the center contact on the lamp through this brass coloured terminal screw

Metal screw shell located in the light fixture

Metal base on the bulb

BATHROOM LIGHT SWITCH

Rules 10-400, 30-200 - Light Fixtures - may be pendant type, such as swag lamps, provided that all the metal on the fixture is properly grounded.

The metal chain on chain-hung fixtures may not be used to ground the fixture - Look for fixtures which have a separate grounding conductor run (threaded) through the chain to each light socket.

All light switches must be kept out of reach of a person in a tub or shower stall. Rule 30-320(3) clearly requires 1 m (39.4 in.) horizontal distance between the switch and a bathtub or shower stall. Although the rule does not say this the wording in Appendix B to this rule suggests that this is a horizontal measurement and that it is measured from the "outside edge" of the bathtub as shown.

Where the 1 m (39.4 in.) distance between the light switch and the bathtub or a shower stall is not possible it may be reduced to no less than 50 cm (19.67 in.) but in that case, the lighting circuit must be GFCI protected.

Heat Lamps - Rules 30-200, 62-110(1)(2) - Like any other recessed light fixture, bathroom heat lamps can be a very real fire hazard if improperly installed. See p. 64 for details on recessed light fixture installation for a heat lamp.

Care should be taken to locate the fixture so that it cannot radiate heat directly onto the upper edge of the door when it is left in the open position, see p. 64 . The rules do not specify any distance, however, a safe horizontal distance may be at least 12 in. from the heat lamp to both the door and shower rod. The reason for this is that any clothes or towels left hanging on the door or on a shower rod may be too close to the fixture and could become overheated and cause a fire.

This distance is measured from the wall of the tub or shower to the switch. It is a horizontal measurement

Wall switch

Tub or shower

FLUORESCENT FIXTURES

Rule 30-310 - Where this type of fixture is mounted end to end in a continuous row as in valance lighting the loomex cable (NMD90) may enter only the first fixture. It must enter the fixture so that it need not run past the ballast. The connection from there to the other fixtures must be made with type A-18, GTF, R90 or similar types of wire. Care should be taken to see that all the fixtures in the row are properly grounded both for safety and for satisfactory operation.

BASEMENT LIGHTING FIXTURES

Rule 30-604 - Pull chain type light fixtures are no longer permitted in wet or damp locations such as near laundry tubs, plumbing fixtures, steam pipes, or other grounded metal work or grounded surfaces in basement or similar areas unless the fixture is approved and marked for use in wet locations. Light fixtures in these locations must be controlled by a wall switch.

LIGHT FIXTURES ON LOW CEILING

Minimum Height of Lighting Fixture - Rule 30-314 - Where fixtures are installed in a crawl space or attic, where there is less than 2.1 m (82.7 in.) headroom, the fixture shall be flexible type or be guarded.

CLOSET FIXTURES

Rule 30-204(2) - Fixtures may not be of the pendant, (suspended) type, nor of the exposed, bare lamp type. They should be located away from any possible contact with stored items in the closet. See the illustration on p. 64.

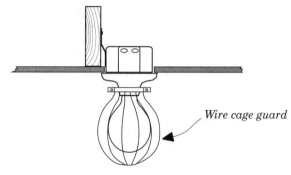

Wire cage guard

BONDING OF LIGHT FIXTURES

Rule 10-808(2) - The light fixture is usually bonded to ground through its own mounting. Where that is not the case it will have a green or bare bonding conductor which must then be connected to the second bonding screw in the outlet box. It may not be placed in the twist-on wire connector with the other bonding conductors in the box because that would require opening the bonding connection to other loads downstream whenever the fixture is removed.

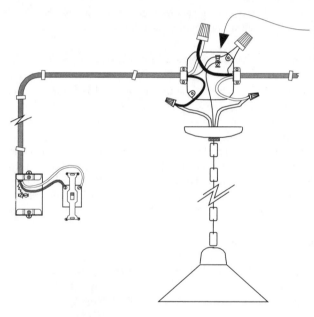

The bonding conductors from the two cables entering the box are connected as shown. The bonding conductor from the fixture connects to the other bonding screw in the outlet box

21. PLUG RECEPTACLES

SET FLUSH

Rule 12-3018 - Check first if the boxes worked out flush with the wall or ceiling finish. Check especially the outlets in feature walls. If they are not flush, see "Box Extenders," p. 98.

POLARIZATION OF PLUG RECEPTACLES

This is an important detail. You will notice the receptacle has a brass terminal screw and a chrome plated terminal screw. Be sure to connect the black or sometimes the red wire to the brass terminal screw and the white neutral conductor connects to the chrome plated terminal screw. The Inspector has a little tester they use to check this connection without removing any cover, etc. If you have connected incorrectly they will find it.

TYPES

Rule 26-700(2) and Diagram 1 in the Code.

Polarized type receptacles must be used for all plug outlets except clock outlets.

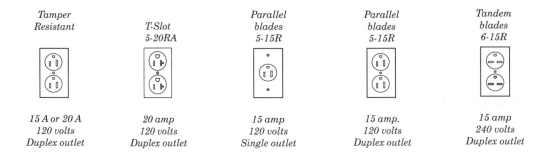

Tamper Resistant	*T-Slot 5-20RA*	*Parallel blades 5-15R*	*Parallel blades 5-15R*	*Tandem blades 6-15R*
15 A or 20 A *120 volts* *Duplex outlet*	*20 amp* *120 volts* *Duplex outlet*	*15 amp* *120 volts* *Single outlet*	*15 amp.* *120 volts* *Duplex outlet*	*15 amp* *240 volts* *Duplex outlet*

Note the difference - The first four receptacles on the left are in common use today. The fifth can be used (should be used) for connection of tankless water heaters, hydro massage bathtub pump motors, ventilating fan motors and central vacuum systems etc. Most of these are supplied with a separate circuit, which may not also supply other loads. The real advantage is safety in maintenance; the appliance can be unplugged and removed for repairs without any concern about having a poor reconnection the appliance is simply plugged in. The last one, a 240 volt receptacle, might be needed in a home workshop.

The T-Slot receptacle - These are 20 amp receptacles and they must be wired with 20 amp conductors and 20 amp breaker but they can be used just like the 15 amp parallel blade receptacles shown above. These will take both the 15 amp parallel blade attachment cap and the 20 amp right angle blade, they are interchangeable now. The T-Slot receptacle must be wired with 20 amp wire and 20 amp breaker at 120 volts.

15 amp. 240 volt plug receptacles are non-interchangeable with the 120 volt receptacles. Single pole circuit breakers used to supply these outlets must be equipped with a tie-bar connecting their operating handles together or use a two-pole breaker. Fuses, if you are using a fuse panel, must have switch or common pull arrangement as used for water heater or dryer, Rule 14-010, 14-302.

Split Receptacle - Note the kind of connection and the type of circuit breakers required for split receptacles. See p. 37, and p. 82.

Tamper resistant receptacles - This is the newest kid on the block. This receptacle automatically blocks off the contact slots when the receptacle is not being used. This safety feature is designed to prevent small metal objects being inserted into the slot and thus touch the live 120 volt internal contact. These are marked Tamper Resistant but on some designs that marking is hard to read. See p. 74 for details. This receptacle is described on p. 52.

GFCI type receptacles - Where GFCI protection is provided in the first receptacle on a kitchen counter plug circuit the connections must be very carefully made. The termination points on these receptacles are clearly marked. The cable from the panel to the first receptacle must be connected to the terminals marked "line." The cable feeding other receptacles downstream must connect to the "load" terminals on that first receptacle. With this connection the second receptacle on this circuit is GFCI protected with the GFCI protection provided in the first receptacle. A standard 20 amp T-slot receptacle is acceptable in the second outlet on that circuit.

Perhaps the best arrangement is to use GFCI type circuit breakers in the panel it simplifies the connection concern and it solves much of the box fill problems but it may be annoying when that GFCI breaker trips and you need to find the panel, then find the tripped breaker to reset it. That cup of tea or coffee at one in the morning may become a bigger project that you counted on. If you plan on locating your GFCI protection for the kitchen counter plugs in the panel, plan ahead. You may not be able to change that arrangement later by simply replacing the standard T-slot receptacle with a GFCI type receptacle, unless you use counter plug outlet boxes properly sized for GFCI type receptacles. These boxes would obviously be deeper than you require for now, but would permit the rearrangement of your GFCI protection later that afternoon. GFCI breakers will trip with only 0.005 amp fault current. Be careful, even a bad thought could make it trip.

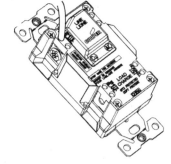

GFCI protection for outdoor plug outlets should be with GFCI type circuit breakers in the panel rather than with GFCI type receptacles. This may prevent some nuisance tripping.

Split Receptacle - These circuits are still acceptable for receptacles on the kitchen counter. Note the kind of circuit breakers required for split receptacles located in the GFCI zone. See p. 82.

GROUNDING AND BONDING

Rules 10-808(2), 10-906 - The bare wire in each outlet box connects first to the box, then to the plug receptacle in every case as shown. In the case of sectional metal boxes the bare wire must connect to the bonding terminal in each section. Note that a pigtail is required so that the plug can be disconnected without opening the bonding connection to other outlets downstream.

BATHROOM PLUG RECEPTACLE

Rule 26-700(11) - Bathroom plug outlets must be either:

• GFCI type receptacle; or
• A standard type duplex plug receptacle provided it is supplied with a GFCI type circuit breaker at the panel. This breaker is called a Class A Ground Fault Circuit Interrupter.

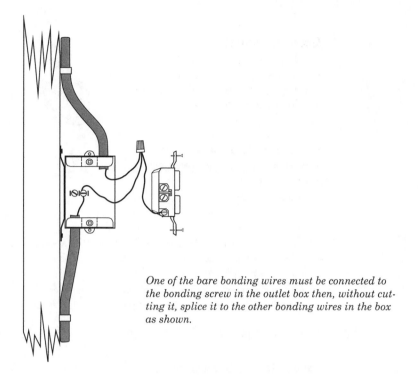

One of the bare bonding wires must be connected to the bonding screw in the outlet box then, without cutting it, splice it to the other bonding wires in the box as shown.

OUTDOOR PLUG RECEPTACLES

Rules 26-724(a), 12-3020, 26-702 - Receptacles installed outdoors, exposed to the weather must be equipped with weatherproof cover plates to prevent moisture from entering. Standard indoor type receptacle cover plates are not acceptable. Instead, wet location cover plates must be used.

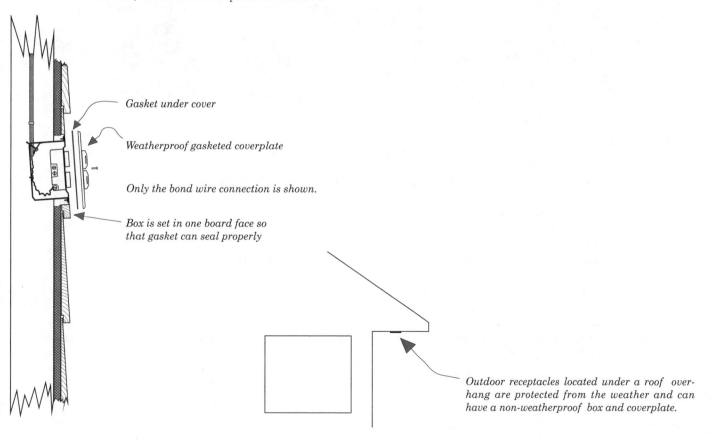

Gasket under cover

Weatherproof gasketed coverplate

Only the bond wire connection is shown.

Box is set in one board face so that gasket can seal properly

Outdoor receptacles located under a roof overhang are protected from the weather and can have a non-weatherproof box and coverplate.

If the outdoor receptacle is installed in a horizontal plane the hot terminals should face down. This may prevent some nuisance tripping due to moisture gathering on the upper surface.

GARDEN OUTLETS

Where an outdoor outlet box is surface mounted on an exterior wall of a building, or is free standing as in a garden area for decorative lighting, the outlet must:

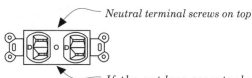
Neutral terminal screws on top

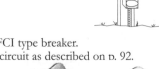

If the outdoor receptacle is installed in a horizontal plane the hot (gold coloured) terminal screws must face downward

- Be in a weatherproof box.
- Be equipped with cover plates held in place with 4 screws.
- Be very well grounded.
- Be above grade.
- These plug outlets must be GFCI type or use a standard receptacle supplied with a GFCI type breaker. These are outdoor plug outlets, therefore they may be supplied with the outdoor plug circuit as described on p. 92.

The supply cable may be direct buried NMWU #14 depending on the length of run. Depth of burial and mechanical protection must be provided as shown on p. 126 for a similar installation.

Rigid steel conduit may be used to protect the cable where it is exposed above ground. Drive a wood or metal post into the ground to support the outlet box.

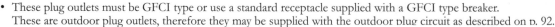

22.   HEAVY APPLIANCES

RANGES, DRYERS, GARBURATORS, ETC.

System Capacity - Rule 8-108 - Before installing additional electrical loads, such as a range, furnace, etc. to a new or an existing service, make sure there is sufficient space in the panel for the branch circuit breakers required to serve the new load. Connecting two or more branch circuits to one fuse or breaker is not approved. Often it is necessary to replace only the branch circuit panel, not the service itself. In any case, the current carrying capacity of the service conductors must, in no case, be less than 60 amperes, Rule 8-200(1)(b)(ii). See also "Table of Service Sizes," p. 18.

Before you begin any electrical work make certain the circuit you will be working on is, in fact, not energized.

Control - Rule 14-010(b) - requires that all 240 volt appliances, such as a range etc. must be provided with a device which will simultaneously disconnect both of the hot conductors at the point of supply.

Circuit Breakers require a tie-bar to fulfill this requirement. Rule 14-302(b)(1).

Length of Run - Any length up to 30 m (98 ft.) is usually acceptable for heavy appliances. Longer runs may be acceptable too but they suggest there is a problem with the service location.

Mechanical protection - Rule 12-518 - requires mechanical protection for loomex cable where it is run on the surface of a wall etc. and within 1.5 m (59 in.) from the floor. To comply with this rule, most Inspectors require a flexible conduit installed over the loomex cable supplying a furnace, garburator, etc.

Cable Strapping - Rule 12-510 - Cable must be properly strapped within 30 cm (12 in.) of cable termination and every 1.5 m (59 in.).

Staples and Straps - used to support cables must be approved for the particular cable involved.

Where cables run through holes in studs or joists they are considered properly strapped. See "Cable Strapping," p. 56 for details.

Bonding - Rules 10-400, 10-404 - All equipment must be adequately bonded to ground. The bare wire in the supply cable must connect to the bonding screw in the branch circuit panel and to each appliance, using the bonding screw or the bolt provided. It is not enough to wrap the wire around a cable connector or cover screw. Too much depends on a good connection.

GARBURATOR

Rules 28-106(1), 28-200, 28-600 - You will require a separate two wire cable to supply the garburator. A #14 copper cable is adequate in most cases.

The supply cable should run into the control switch outlet box above the counter then down to the motor. You will require flexible conduit to protect the cable from a point inside the wall to the connector on the garburator.

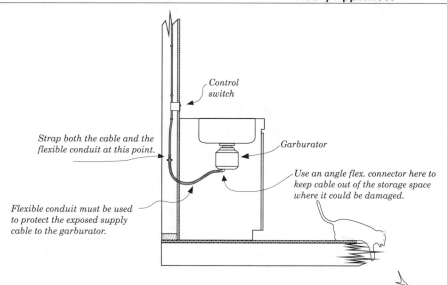

Control switch

Strap both the cable and the flexible conduit at this point.

Garburator

Use an angle flex. connector here to keep cable out of the storage space where it could be damaged.

Flexible conduit must be used to protect the exposed supply cable to the garburator.

DISHWASHER

Rules 28-106(1), 28-200 - This is a motor and heating load. These usually operate at 120 volts, therefore, a 2-wire cable is required. Unless you are absolutely sure your dishwasher can be served with a #14 cable you should install a #12 copper and 20 amp breaker. You will require a separate circuit. This cable may be run directly into the connection box on the dishwasher. Make sure that the bare wire is connected to the bonding terminal.

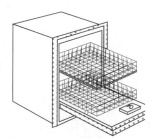

DOMESTIC RANGES

Rules 8-300, 26-744(4), (and in Ontario, Bulletins 26-14-4)

FREE STANDING RANGE

Rules 8-300; 26-744; 26-746 - Freestanding electric ranges must be cord connected. The plug receptacle required for this connection is a 3 pole, 4 wire grounding type as shown below.

Note 1: This applies in every case. If for any reason a range is being replaced with a new or another used one, it must then be cord connected.

Note 2: If provision is made for a free standing range, electric or gas, the rule requires that a plug outlet be installed for a standard electric range. This is in addition to the piping for the gas range. See p. 89 for details.

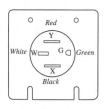

Red
Y
White *W* *G* *Green*
X
Black

Range receptacle

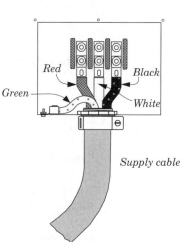

Red   *Black*
Green
White

Supply cable

Cable size ....#8 NMD90 copper
Outlet box size ....4 11/16 x 4 11/16 x 2 1/8
Plug receptacle rating ....45 amp.
Rating of fuses or breakers ....50 amp each

Rule 12-3012(3) says this outlet box must be fastened to a solid member directly behind the box or be supported on two of its sides. If possible locate the box in a stud space where it can be secured to both the plate and a stud, or if both sides of the wall will be finished the box may be supported as shown below.

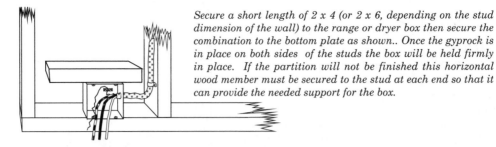

Secure a short length of 2 x 4 (or 2 x 6, depending on the stud dimension of the wall) to the range or dryer box then secure the combination to the bottom plate as shown.. Once the gyprock is in place on both sides of the studs the box will be held firmly in place. If the partition will not be finished this horizontal wood member must be secured to the stud at each end so that it can provide the needed support for the box.

There are 3 rules to watch for:
• The range outlet box must be located very near the mid-point on the wall behind the range.
• This range outlet box must not be higher than 13 cm (5.1 in.) above the floor to the center of the outlet box.
• This range outlet box must be carefully positioned so that when the receptacle is finally installed, the ground pin will be either on the right hand or the left hand but not at the top or the bottom.
• Flush mounted - Where practicable, the range plug outlet box must be flush mounted, Rule 26-744(9). This is to provide more storage space for the supply cord and to allow the range to move fully back against the wall.

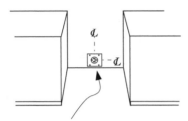

Set box 5.1 inches (130 mm) floor to centre of the box. The ground pin must be in either the 3 o'clock or 9 o'clock position but may not be in the 12 o'clock or 6 o'clock positions. Make sure the front cover mounting screws, on the box, are in the correct position to allow the receptacle to be mounted properly.

DROP-IN RANGE

This is a conventional range but it is not free standing. It is fitted into the kitchen cabinets.
Cable size .... #8 NMD90 copper
Outlet box size .... box not needed
Flexible conduit size (if required) .... 3/4 in.
Rating of fuses or breakers .... 50 amp
This unit is not cord connected. The #8 NMD90 cable may be run directly into the connection box.

The supply cable must be protected with 3/4 in. flexible conduit for the last 3 ft. or so at the range if it is subject to mechanical damage.

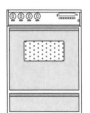

BUILT-IN RANGE

Separate units - Rule 26-744
Main cable size .... #8 NMD 90 copper
Flexible conduit size (if required) .... 3/4 in.
Cable to oven size .... #10 NMD90
Flexible Conduit size (if required) .... 1/2 in.
Rating of fuses or breakers .... 50 amp each

The above sizes are sufficient for the average 12 kw range

Separate Built-in Cooking Units - Rule 26-746(3) says that the two units, the cooking top and the oven, "shall be considered as one appliance." See also Rule 8-300(2). Then notice that Rule 26-744(1) says that an electric heating appliance "shall have only one point of connection for supply." These two rules therefore, will not permit us to run a separate supply cable from the panel to each unit.

Rule 26-742 was revised again to limit the #10 copper taps from the junction box to the oven and the cooking tops, to 25 ft each, as noted in the illustration below. If a longer tap is needed then it must be the same size as the main range supply cable, in most cases this is #8 copper. This change is consistent with other rules which allow a smaller tap without overcurrent protection for certain loads.

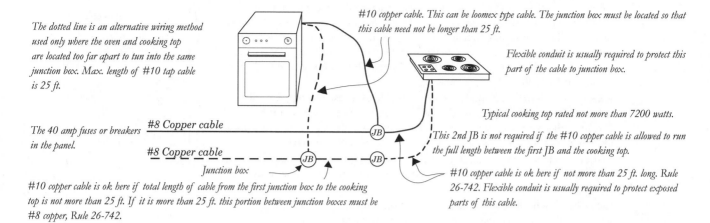

The dotted line is an alternative wiring method used only where the oven and cooking top are located too far apart to tun into the same junction box. Max. length of #10 tap cable is 25 ft.

#10 copper cable. This can be loomex type cable. The junction box must be located so that this cable need not be longer than 25 ft.

Flexible conduit is usually required to protect this part of the cable to junction box.

The 40 amp fuses or breakers in the panel.

#8 Copper cable

#8 Copper cable

Junction box

Typical cooking top rated not more than 7200 watts.

This 2nd JB is not required if the #10 copper cable is allowed to run the full length between the first JB and the cooking top.

#10 copper cable is ok here if total length of cable from the first junction box to the cooking top is not more than 25 ft. If it is more than 25 ft. this portion between junction boxes must be #8 copper, Rule 26-742.

#10 copper cable is ok here if not more than 25 ft. long. Rule 26-742. Flexible conduit is usually required to protect exposed parts of this cable.

It should also be noted that the manufacturer of electric ranges is not required to provide overcurrent protection for the individual elements on the cooking top, or the oven, nor are we, as installers, required to provide that overcurrent protection when we install these units. Ranges which have a plug outlet for use with smaller appliances will have a 15 amp fuse or breaker protecting that circuit only. The only protection provided for the range elements is the 40 amp breaker in the supply panel.

Cable protection - Flexible conduit is usually required on these cables for mechanical protection where they are subject to mechanical damage, even though they are in a cabinet below the range, Rule 12-518.

Grounding - If you do not connect the bare ground conductor properly at the panel and at the range, one day the chief cook may not be alive to greet you with a kiss at the end of a busy day.

Gas Range

Rule 26-712(d) says that if the piping for a free-standing gas range is installed then we must also install a 15 amp plug outlet in that space for the range. This plug outlet must be located the same as the 50 amp plug for an electric range as shown on p. 115. This range plug may be supplied from the kitchen eating area plug circuit as shown on p. 89.

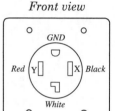

Front view

30 amp Dryer receptacle

Dryers

Rules 26-744(2)&(3) - Electrical dryers must be cord connected, they may not be hard wired. Install a 3 pole, 4 wire grounding type plug receptacle behind the dryer location.

If for any reason a hard wired dryer is being replaced with a new or another used one, the rules require a plug outlet be installed at the dryer location.

Cable size .... #10 NMD90 copper
Outlet box size .... 4-11/16 x 4-11/16 x 2-1/8 in.
Plug receptacle rating ....30 amp.
Rating of fuses or breakers .... 30 amp each

These sizes are sufficient for most dryers, i.e. dryers with ratings not greater than 7200 watts.

Rule 12-3012(3) says this large box required for the dryer receptacle must be fastened to a solid structural member directly behind the box or on two of its sides.

The bare grounding conductor must be properly connected at the panel and at the dryer to ensure safety to the operator.

Water Heater

Circuit - Rule 26-750(4) - requires a separate circuit with sufficient capacity to carry the maximum possible load that may be connected at one time by the thermostat. Most tanks are provided with 2 - 3 kw. elements, which totals 6,000 watts, however, the internal thermostat control allows only one of these elements to operate at one time. See Notes 8 & 9 on p. 11 for details on this control arrangement.

Cable size ....
#12 NMD90 copper
Flexible conduit size .... 7/16 in.
Rating of fuses or breakers .... 20 amp.
Fuse must be Type P or D ....Rule 14-610

These sizes refer to water heaters with ratings from 3 kw. (3,000 watts) up to 3.8 kw. (3,800 watts).

According to Rule 8-302(2) a water heater is usually considered a continuous load. In that case the supply circuit must not be loaded to more than 80% of the rating of the breaker or fuse supplying it. A 3 kw. water heater draws 12.5 amps at 240 volts but the largest continuous load permitted on a 15 amp circuit is 80% or 12 amp. We therefore require 12.5 x 10/8 = 15.6 amp. This means we need a 20 amp breaker or fuse and #12 copper cable.

Rule 4-034(1) permits a 2-conductor loomex cable to be used to supply a 240 volt load. When this is done the white wire is not used as a neutral but is hot. Wherever this white wire is visible, as in a junction box etc. it must be painted with black non-metallic paint, or better still, just wrap it with black electrical tape.

Location of Hot Water Tank - Rules 2-118 & 26-750(3) Locate the tank carefully; the supply connections, service covers and nameplate data must all be accessible after completion of the building.

Cable Protection - Rules 12-518, 12-1004 - Protect the hot water tank supply cable with 7/16 in. flexible conduit where exposed to mechanical injury.

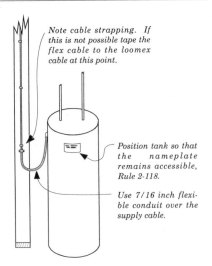

Note cable strapping. If this is not possible tape the flex cable to the loomex cable at this point.

Position tank so that the nameplate remains accessible, Rule 2-118.

Use 7/16 inch flexible conduit over the supply cable.

FURNACE (GAS AND OIL)

RULE 26-806(1)

Some have interpreted this rule to mean that the supply cable to the furnace must not be with a 3-wire cable, but that it must be a separate two wire cable all the way back to the panel. To avoid a problem it is best to run a separate 2-wire cable to the furnace, or check with your local inspector.

GAS FURNACE

Cable size ....#14 copper
Flexible conduit size..7/16 in.
Fuse or breaker size..15 amp

OIL FURNACE

Cable size ....#12 copper
Flexible Conduit size..7/16 in.
Fuse or breaker size..20 amp

Disconnect Switch - Rules 26-806(5)(6)(7), 28-600 - A disconnect switch is required for each furnace.

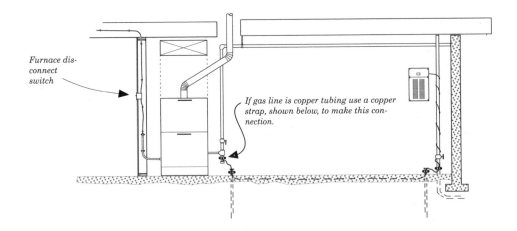

Furnace disconnect switch

If gas line is copper tubing use a copper strap, shown below, to make this connection.

Location - A circuit breaker in a branch circuit panel may serve as disconnect switch provided it is located between the furnace and the escape route.

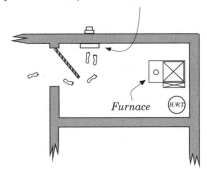

In this case the circuit breaker in the panel is accessible without having to pass the furnace. This is an acceptable location for the disconnect switch

Furnace

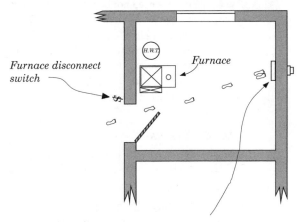

Furnace disconnect switch

Furnace

The branch circuit breaker in this panel may not serve as the furnace disconnect switch because the escape route is past the furnace. The furnace disconnect switch must be located as shown.

Protection - Rule 26-806 - The loomex supply cable should be run in a 7/16 in. flexible conduit for the last few ft. near the furnace where it is less than 1.5 m (59 in.) above the floor or where it is subject to mechanical damage.

Bonding

Rule 10-400 - Be sure to connect securely the bare bonding conductor in the supply cable to the bonding terminal in the furnace connection box.

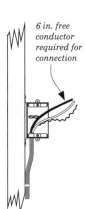

6 in. free conductor required for connection

Gas Piping

Rule 10-406(4) - All interior metal gas piping used to supply the furnace, fire place, or other appliance, must be bonded to the nearest grounded cold water pipe or to the service grounding electrode as shown in the illustration above. In that illustration the service grounding conductor continues on from the last ground rod to the gas line at the furnace. Do not cut the grounding conductor, simply run it through the ground clamps and then on to the last clamp on the gas line.

The gas lines used today are not like the rigid steel pipe we used to see in the good old days. Gas lines used today are of copper and in some cases of soft copper. Do not use an ordinary ground clamp on any copper gas lines because it may damage the pipe. Rule 10-614(2) allows us to use a copper strap to make this bonding connection. See illustration on the left.

Kitchen Fan Range Hood Fan, Ventilation Fan, Bathroom Fan, Rotary Fan, Attic Fan

Rule 28-106 - These fan motors are often shaded pole type drawing less than 4 amps each. Many of these motors draw less than one amp and in that case they are considered as any general purpose lighting or general purpose plug outlet. A 15 ampere circuit can supply a mixture of 12 such outlets. These fan motors are normally supplied with a general lighting or general plug circuit. Technically, however, where the motor draws more that one amp, subtract that amp load from the maximum 12 amp load permissible for a normal 15 amp circuit. The remaining ampacity can then be used for other general lighting or general plug outlets at one amp per outlet. For example, a fan motor drawing say 1.5 amps would be counted as the equivalent of 2 outlets, if it draws 2 amps it would count as 2 general outlets, if it draws 3 amps it would be counted as 3 general purpose outlets. Subtracting the load of, say a 3 amp fan load from the maximum 12 amp total load leaves a balance of 9 amps for 9 other general light or general purpose plug outlets which may be connected to that same circuit. The Code allows only 12 such outlets on a 15 amp circuit but that is the maximum load permitted. It is better to have only 9 or 10 so that if you suddenly discover that you need more outlets in that part of the house you can add that outlet without becoming a common criminal by having 12-1/2 outlets on a circuit.

Cable size ....#14 NMD90 copper
Flexible conduit size (If required) ....7/16 in.
Fuse or breaker rating...15 amp.

The Kitchen Fan - need not be on a separate circuit. It may be on a general lighting or plug circuit but it may not be on any special kitchen appliance circuits, Rule 26-722(b)(ii). It is counted as one outlet when determining circuit loading.

Cable Protection - Rule 12-518 - Protect loomex cable with 7/16 in. flexible conduit where it is exposed to mechanical damage. This includes that portion of the cable which is exposed inside the cabinet where it may be subject to damage. Flexible cable must terminate in a flex connector at the fan and be strapped or taped to the loomex cable before it emerges from the wall.

Grounding - Rule 10-400 - Because it is near grounded objects it is very important to connect the bare ground wire securely to the ground terminal in the fan connection box.

VACUUM SYSTEM

Rule 26-710(1) requires that if piping for a central vacuum system is installed a plug outlet must be provided for the unit. Rule 26-722(e) requires that this plug outlet be supplied with a separate circuit used for no other load.

HYDRO MASSAGE BATHTUB

CSA Certified Units - Make sure the unit you install has one of the certification marks illustrated on p. 2. This may seem like a picky point but in fact it is extremely important for your safety.

SEPARATE CIRCUIT REQUIRED

Rule 28-106(1) - This is a motor load, therefore, the rules in Section 28 must be applied. For the rough wiring stage, run a separate 2-wire loomex cable to the pump motor location. The size of this cable is determined by the nameplate ampere rating. Later when the unit is actually installed you will need to install a disconnect switch at the motor location.

DISCONNECT SWITCH REQUIRED

Rules 28-600, 28-602(3)(e) - These rules require a disconnect switch for this motor. This may be a standard wall switch, such as is used to control light outlets. Don't forget, the ampere rating of this switch must be at least 125% of the ampere rating of the pump motor.

Note This is not a hot tub. It is a hydro massage unit only. A hot tub usually has an electric heating element as well as a pump.

One more thing - do not forget to provide access to this motor and its disconnect switch for maintenance purposes. See the illustration below.

GFCI REQUIRED

Rule 68-302 - This motor circuit must be protected with a GFCI type circuit breaker. There is no exception to this rule. Because this is a motor load this special circuit breaker may not also supply any other load.

LIGHTS, SWITCHES, AND PLUGS

Rule 68-304 - The usual rules for lights, switches and plugs in bathrooms also apply to this special bathroom.

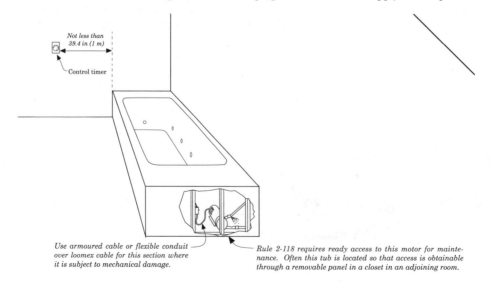

Not less than 39.4 in (1 m)

Control timer

Use armoured cable or flexible conduit over loomex cable for this section where it is subject to mechanical damage.

Rule 2-118 requires ready access to this motor for maintenance. Often this tub is located so that access is obtainable through a removable panel in a closet in an adjoining room.

CONTROL OF PUMP MOTOR

Rule 68-304(2) - Timer is not Required - Rule 68-304(1) - The rule now requires only a simple wall switch properly rated (125% of motor full load amps) to control this pump motor. A timer switch may be used if you want that kind of control.

LOCATION

Rule 68-304(2) - The control switch must be at least 1 m. (39.4 in.) away from the tub. This is a horizontal measurement from the outside wall of the tub, (the rule says "from the wall of the hydro massage tub") to the control switch. There is an exception - where the control switch is an integral part of a properly certified factory built hydro massage bathtub it may be closer than 1 m from the tub. In fact, in that case, the control switch may be easily accessible to anyone in the tub. Do these factory people know something the rest of us do not know or do they just use better switches? Maybe they use better bonding methods or maybe they just use better arguments. Just kidding, that control system is safe.

23.   ELECTRIC BASEBOARD HEATING

HEAT LOSS CALCULATION

Check if your Building Code or local electric power utility requires a heat loss calculation. This is a special method of calculating heat loss based on the type of construction, the quality of building insulation and the geographical location of the house. This is a fairly accurate method of determining the size of heating elements needed in each room. If a heat loss calculation is required the organization requiring it will normally advise you on the calculation methods they want you to use and will provide you with the necessary forms.

ROUGH FLOOR MAP REQUIRED

The Electrical Inspector will want a rough sketch of the floor plan of your house showing:
* The location of heaters
* The rating of each heater
* What circuit it is on
* Size of supply conductors used
* The rating of breakers

Be sure your sketch is accurate and clear - easy to follow. Your Inspector will want it for that first rough inspection. See p. 59 for details on drawing a floor plan and how to identify each circuit.

BRANCH CIRCUITS

Rule 62-110 - Branch circuits which supply electric heaters may not be used to supply any other load.

2-Wire Cables - Electric heaters are usually connected for 240 volts - no connection to the neutral. Rule 4-034(1) permits a 2-wire loomex cable with one black and one white wire to be used for these loads. Fact is, that white wire is operating under false pretenses, it is not a neutral in such cases and it needs to be properly introduced as a live, hot conductor. A little black, or red tape, on the exposed sections of this white conductor will solve this problem.

BREAKERS

Tie-bars - Rule 14-302(b)(i) - Two single breakers require a tie-bar when used to supply 240 volt appliances such as heaters. The tie-bar is used to mechanically connect the operating handles of the two breakers so that they operate as one.

CIRCUIT LOADING

Rule 62-114(7) & (8) - The rules regarding electric heating were changed in the 1990 Code. Since then we have been allowed to use the full rated ampacity of the cables supplying electric baseboard heating. For example, a #14 copper conductor may carry 15 amps at 240 volts and therefore is allowed to supply 3600 watts of heating load. To make this possible Rule 62-114(7) was changed so that for electric heating branch circuits only, the supply breaker for a #14 copper cable could be 20 amp, and a 25 amp breaker could be used to supply a #12 copper branch circuit heating cable as shown in the illustrations.

This change in the rules was based on two very important factors. First, the circuit breaker in this heating circuit is not needed to limit the load on these conductors, it is there to automatically open the circuit in the case of a short circuit and, of course, for circuit maintenance purposes. Please note that unsafe and illegal additional heaters are easily connected to these circuits but that is true for every circuit in the house. Second, the fact that fixed heating loads are just that, they are fixed, they do not vary, they are either on or off. This is in contrast to other branch circuits in the house which supply plug outlets or lighting outlets where the total load is constantly changing and is somewhat unpredictable. The following two illustrations show circuit loadings permissible under the changed rules.

20 AMP CIRCUIT BREAKER, #14 COPPER CABLE, MAXIMUM LOAD PERMITTED IS 3,600 WATTS

The actual wattage rating and the number of heaters used need not be as shown provided the sum of the rating of all the fixed heaters on the circuit does not exceed the maximum permitted for that circuit.

See also "Notes for the Student - Overcurrent Protection," p. 124 for Electrical Code verification.

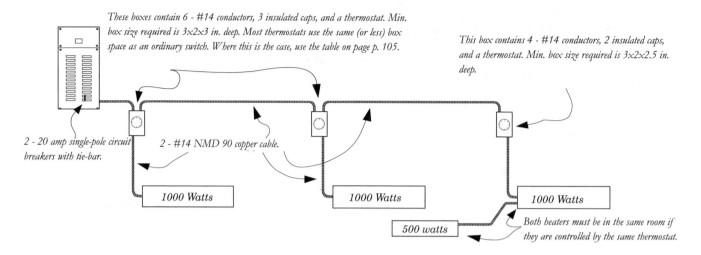

These boxes contain 6 - #14 conductors, 3 insulated caps, and a thermostat. Min. box size required is 3x2x3 in. deep. Most thermostats use the same (or less) box space as an ordinary switch. Where this is the case, use the table on page p. 105.

This box contains 4 - #14 conductors, 2 insulated caps, and a thermostat. Min. box size required is 3x2x2.5 in. deep.

2 - 20 amp single-pole circuit breakers with tie-bar.

2 - #14 NMD 90 copper cable.

1000 Watts

1000 Watts

1000 Watts

500 watts

Both heaters must be in the same room if they are controlled by the same thermostat.

Maximum Length - Maximum length of supply cable to the first heater in the circuit should not exceed 82 ft. (Approx. 25 m). After this first heater the load is smaller and length of run is not usually a problem.

The maximum load permitted with #14 copper cable must not exceed 15 amp. The 20 amp breaker shown protecting this cable is permitted because this is a fixed load. Maximum circuit load, (in watts) must not exceed conductor ampacity multiplied by the circuit voltage. For example, a 15 amp conductor operating at 240 volts could supply 3600 watts of electric baseboard heating.

Do not bundle these cables, maintain separation as described on p. 53.

Under the old rules we would not have been permitted to use a 20 amp breaker; 15 amp was the max rated breaker allowed for a #14 copper cable. With that old arrangement the maximum load permitted was only 2880 watts.

25 AMP CIRCUIT BREAKER, #12 COPPER CABLE, MAXIMUM LOAD PERMITTED IS 4,800 WATTS

See also "Notes for the Student - Overcurrent Protection," p. 124 for Electrical Code verification.

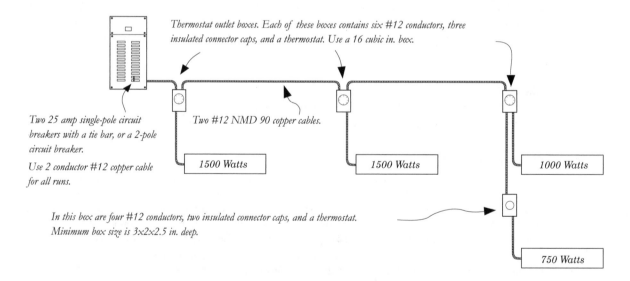

Thermostat outlet boxes. Each of these boxes contains six #12 conductors, three insulated connector caps, and a thermostat. Use a 16 cubic in. box.

Two 25 amp single-pole circuit breakers with a tie bar, or a 2-pole circuit breaker.

Two #12 NMD 90 copper cables.

Use 2 conductor #12 copper cable for all runs.

1500 Watts

1500 Watts

1000 Watts

In this box are four #12 conductors, two insulated connector caps, and a thermostat. Minimum box size is 3x2x2.5 in. deep.

750 Watts

Maximum Length - Maximum length of supply cable to the first heater in this circuit should not exceed 100 ft.(Approx. 30 m). After the first heater the load is smaller and length of run is not usually a problem.

The maximum load permitted with #12 copper cable must not exceed 20 amp. The 25 amp breakers shown protecting this cable are permitted because it is a fixed load. Given the above configuration, one could also use a preferred 2-pole circuit breaker. Maximum circuit load must not exceed conductor ampacity multiplied by the circuit voltage. In this case it is 20 amps x 240 volts = 4800 watts.

Note - In the above illustration Rule 62-114(4) will permit #14 copper cable to be used between the thermostat and the heater provided that cable length is not more than 24.6 ft (7.5 m). This distance is not measured as the crow flies, but is actual cable length. That maximum length of 24.6 ft. worked just fine about 30 years ago when we built smaller houses, but in today's much larger houses the rooms are larger, and the 24.6 ft. maximum length is simply not enough in many cases. Any taps longer than 7.5 m must be made with #12 copper cable.

It is still permissible to use a 15 amp breaker and #14 copper cable to supply 2880 watts or a 20 amp breaker on #12 copper cable supplying 3880 watts but there is no need to do so now. It should be noted that the rules always did permit the cables to carry their full rated current, the trouble was the restriction placed on the breaker supplying the cable. It could only be loaded to 80% of its rating. That has not changed either. What has changed, to make this greater loading possible, is Rule 62-114(8) which permits higher rated breakers for the #14 and #12 cables when they supply fixed heating loads.

Do not bundle these cables, maintain separation as described on p. 53.

THERMOSTATS

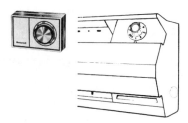

Location - Rule 62-202 requires a thermostat in each room where electric heating is installed. All the heaters in a room must be controlled by a thermostat located in that room.

Location in Bathrooms - The rules do not specifically require the control for electric heaters in a bathroom to be at least 1 m (39.4 in.) horizontal distance from the tub or shower stall, however, that is better. It should not be possible to operate the thermostat from any position in the tub. Watch this in the rough wiring stage - make provision for the thermostat.

Rating - Rule 62-118(1) - Those thermostats which are connected directly into the line and control the full load current must have a current rating at least equal to the sum of the current ratings of all the electric heaters they control.

Type - A single pole thermostat which does not have a marked "off" position is acceptable on a normal 240 volt heating circuit.

Background - Rule 62-118(2) accepts thermostats whether they are marked with an indicating "off" position or not, it does not matter. This sub-rule merely indicates that if a thermostat has a marked "off" position that it must then open, (in that case only) both hot conductors of the controlled heating circuit. If the thermostat only indicates a high and a low position with graduated markings between, it need not open all ungrounded conductors of the circuit. This means that the heating circuit is always hot (energized) but open until the thermostat calls for heat.

OUTLET BOXES

Rules 12-506, 12-3002(6) - Location - An outlet box may be installed behind the heater but that is not the best way to make the connection. This is perhaps the poorest method of connection to use because the box would need to be very accurately located to ensure it will be behind the heater and that it will be properly covered when the heater is installed. Bonding is a problem with this arrangement and finally, access to splices in the box is unsatisfactory. Avoid this method if possible. It is better to run these cables into the thermostat outlet boxes as shown in the circuit illustrations above. Then, in that case, only one cable need enter the heater and this should run directly into the heater terminal box.

The double cable connector in some heaters is an advantage when there are two or more heaters in the same room as shown in the illustration on p. 122.

CABLE PROTECTION

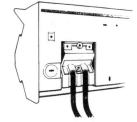

Rule 12-518 - Where part of the cable is run exposed to mechanical injury, use a short length of EMT conduit or flexible conduit to protect the cable. Be sure to terminate the metal conduit in an approved manner so that it is grounded.

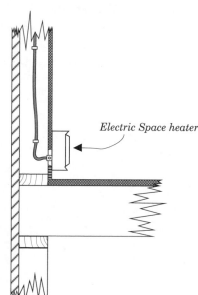

Electric Space heater

BATHROOM HEAT LAMP

Rule 62-108(3) - Heat lamps in bathrooms are normally supplied from general use lighting or plug outlet circuits. They may not be supplied from electric heating circuits. The reason is that the heat lamp is not the only heating provided in a bathroom, it is a supplementary heat source for a specific purpose. The control switch for a heat lamp in a bathroom must be at least 1 m (39.4 in.) from the outside face of a tub or shower. This is illustrated on p. 67.

INFRARED RADIANT HEATERS

Where these are the metal-sheath glowing element type, ground fault protection should be installed to de-energize the normally ungrounded conductors. Where there are multiple heaters on the same branch circuit, a single means of ground fault protection is permitted.

GROUNDING AND BONDING

Rules 10-400, 10-906 - The bare grounding wire (the Code now calls this a bonding wire) in the supply cable must be securely connected to the fixture with the grounding terminal screw in each fixture.

NOTES FOR THE STUDENT - OVERCURRENT PROTECTION

- 20 amp breaker for a #14 copper loomex cable to supply 3,600 watts of electric heating; or
- 25 amp breaker for a #12 copper loomex cable to supply 4,800 watts of electric heating.

Are these loads legal?

Lets look at the actual Rules:

Rule 62-114(7) - Service, feeder, or branch conductors supplying only fixed resistance heating loads shall be permitted to have an ampacity less than the rating of the circuit overcurrent protection, provided that their ampacity is:

      (a.)   not less than the load, and

      (b.)   at least 80% of the rating of the circuit overcurrent protection.

Rule 62-114(8) - Notwithstanding Subrule (7)(b), 125% of the allowable ampacity of a conductor does not correspond to a standard rating of the overcurrent device, the next higher standard rating shall be permitted.

Table 16B - Bonding conductor size problem - The #12 copper loomex cable has a #14 bonding conductor and some have argued that this is too small for a 30 amp breaker. Table 16B is intended to specify the size of the bonding conductor required for an electrical installation, in this case electric baseboard heaters. Please note that Table 16B is based on the ampacity of the circuit conductors in the cable. It is not based on the breaker rating. To find the correct size bonding conductor required for this #12 copper loomex cable we must enter Table 16B at the 20 amp level and read across under the second column marked "Size of bonding conductor." The table requires a #14 copper bonding conductor for the #12 copper loomex cable and that is exactly what the manufacturer put into that cable. This ratio of bonding conductor ampacity to circuit conductor ampacity is normal for almost all electrical installations. Yes, the bonding conductor in the cable is large enough for the 30 amp breaker but, according to Code, we should use the next size larger (30 amp) only if the 25 amp rated breaker is not a standard rating. 25 amp circuit breakers are generally available now and that means we may not use the 30 amp breaker now.

Yes, these loads and these breaker ratings are safe and they are legal.

24.   ELECTRIC HOT AIR FURNACE

RULE 62-208

Installing an electric hot air furnace is not as difficult as it may appear. By carefully following these basic instructions, you can do it and save.

CLEARANCES

Rule 62-208 - This rule does not specify any minimum clearance but it does draw attention to two basic clearances required.

From Combustible Surfaces - Observe all the clearances specified on the furnace name plate and,

For Maintenance - Do not install your furnace in a small confined space unless it is designed and marked for installation in an alcove or closet. There must be sufficient clearance to allow removal of panel covers and for maintenance work.

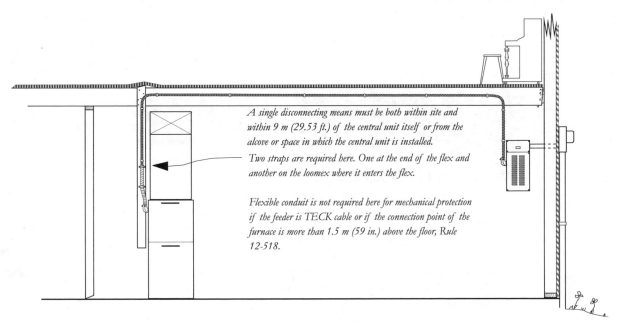

A single disconnecting means must be both within site and within 9 m (29.53 ft.) of the central unit itself or from the alcove or space in which the central unit is installed.

Two straps are required here. One at the end of the flex and another on the loomex where it enters the flex.

Flexible conduit is not required here for mechanical protection if the feeder is TECK cable or if the connection point of the furnace is more than 1.5 m (59 in.) above the floor, Rule 12-518.

SUPPLY TO FURNACE

Rule 62-114(6)&(7) - Below is a greatly simplified table of sizes. Use this table to determine the rating of the furnace supply cable and circuit breakers.

The table also takes into account the current drawn by the fan motor.

Your Furnace Nameplate Rating	Fuse or Circuit Breaker Rating	Size of supply conductor to furnace	
		Using Copper	**Using Aluminum**
5 KW	30 Amp	#10 NMD90 (Loomex Cable)	#8 NMD90 (Loomex Cable)
10 KW	60 Amp	#6 NMD90 (Loomex Cable)	#6 NMD90 (Loomex Cable)
15 KW	100 Amp	#4 NMD90 (Loomex Cable)	#3 NMD90 (Loomex Cable)
20 KW	125 Amp	#3 NMD90 (Loomex Cable)	#2 NMD90 (Loomex Cable)
25 KW	150 Amp	#2 NMD90 (Loomex Cable)	#0 NMD90 (Loomex Cable)
30 KW	175 Amp	#1 NMD90 (Loomex Cable)	#00 NMD90 (Loomex Cable

The above Table is based on Chromalox SPEC. sheet CCSL 266.4. dated Oct. 18/82

FURNACE DISCONNECT SWITCH

A single disconnecting means that simultaneously opens all ungrounded conductors that supply the controller and the central unit are required. Where there are multiple circuits supplying a central unit, the multiple disconnecting means must be grouped and a label must indicate that there are multiple circuits that must be isolated before the central unit can be serviced.

WIRING METHOD

NMD-9O cable. This cable may be used if the size you need is available. All the usual rules for running loomex cable apply to this cable, i.e. strap it, protect it, and terminate it in approved cable connectors.

Teck Cable (copper) may also be used to supply the furnace. This is a tougher cable and it costs a lot more but it may not need any additional protection from mechanical injury.

Dry type cable connectors may be used indoors.

Peel off only the PVC outer sheath, and only at the connector, so that it (the dry type connector) is clamped directly onto the metal armour.

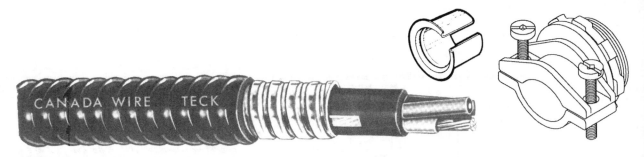

Mechanical Protection Required

Teck Cable has an aluminum armour and can take a little abuse, however, if there is any doubt, provide protection for the cable especially those sections which are run on the surface below the 5 ft. level.

NMD7 or NMD90 - Loomex Cable - has no armour and therefore must always be protected from mechanical damage where it is run on the surface below the 5 ft. level. Rule 12-518. Use a flex conduit, sized large enough to allow the cable to move freely, for those sections where mechanical protection is needed.

This flex conduit must then terminate in a flex connector at the furnace.

Consider using an angle connector at the furnace because this cable, and the flexible conduit, may not be bent sharply without damaging it.

Thermostat Control Wiring

Table 19 in the Code lists only one cable acceptable for low voltage thermostat control work. That is type LVT cable. Truth is, Table 19 does list a type ELC cable for Class 2 wiring but Rule 16-210(3) says it may not be used for heating control (thermostat) circuits. So we are stuck with having to use LVT cable only.

Strapping - Whatever staples or straps you use to support this small cable make sure they do not damage the sheath. Staples should be driven in only till they contact the cable. When a staple is driven in too deeply and short circuits the conductors, it turns the furnace on and there is nothing to shut it off, automatically, except the high limit safety device inside the furnace. The thermostat cable is important - install it carefully.

Grounding

Rules 10-400, 10-600 - Make certain the grounding conductor is properly connected with approved lugs, at each end. Look for a separate grounding lug in the furnace connection box and in the service panel.

25.   Driveway Lighting

Conduit System

Conduit is very seldom used for this work. It is difficult to properly terminate the conduit at either end. Don't forget, a grounding conductor must be drawn into the conduit along with the circuit conductors to ground the light standard.

Polyethylene pipe is usually permitted for this installation provided we use it only for the underground portion and the cable we use is an underground cable such as NMWU or NMW10 loomex cable. The riser at the house must be rigid metal conduit or rigid PVC pipe.

Direct Burial

Rule 12-012 - Depth of Burial - Type NMWU (or the old NMW10) loomex cable, size #14, may be used as shown in the illustration. It must be buried to a depth of 60 cm (23.6 in.) or if it passes under a roadway or driveway, to a depth of 90 cm (35.4 in.). This is based on Code Table 53.

These depths may be reduced by 6 in., (15 cm) if the cable is protected by treated planking as shown in the illustration.

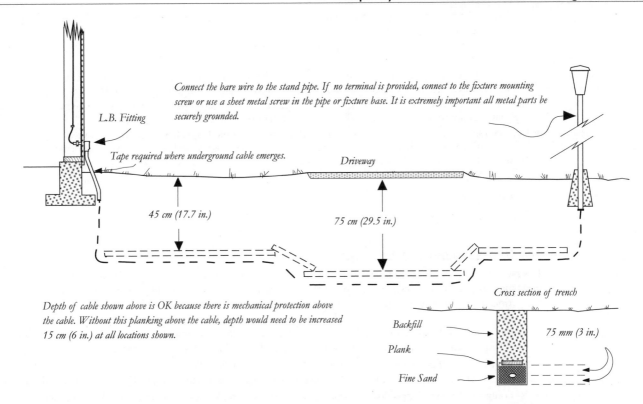

Connect the bare wire to the stand pipe. If no terminal is provided, connect to the fixture mounting screw or use a sheet metal screw in the pipe or fixture base. It is extremely important all metal parts be securely grounded.

L.B. Fitting

Tape required where underground cable emerges.

Driveway

45 cm (17.7 in.)

75 cm (29.5 in.)

Cross section of trench

Backfill

Plank

Fine Sand

75 mm (3 in.)

Depth of cable shown above is OK because there is mechanical protection above the cable. Without this planking above the cable, depth would need to be increased 15 cm (6 in.) at all locations shown.

Sand or earth - may be used but note the stuff is to be screened with #6 mm (1/4 in.) screen. Cables should lie on 75 mm (3 in.) thick bed of this screened sand or earth and have a further 75 mm (3 in.) thick blanket of screened sand placed on top of the cables.

Treated Planking - 2 in. untreated cedar planking is acceptable in most cases. The Rule says a 38 mm (1-1/2 in.) treated plank is required. Check with your Inspector before you use untreated planks.

Section of Conduit - This is to protect the cable from damage. Rule 12-012(5) requires that this conduit terminate in a vertical position approximately 30 cm. (11.8 in.) above the trench floor. The cable must continue on downward, as shown, to allow movement during frost heaving.

Size of Conduit - is not critical. It should be large enough so that cable can be drawn in easily and without damage to the insulation.

Type of Conduit - may be rigid metal conduit or P.V.C. but may not be EMT, Rule 12-1402, 22-500.

GROUNDING AND BONDING

Rules 10-400, 10-906 - It is very important that the ground wire be securely connected to the lamp standard.

26. TEMPORARY CONNECTION TO PERMANENT BUILDING

SECTION 76

This is really a permanent service connection but it is on a temporary basis until the wiring is entirely completed. Many of the building contractors ask for this connection to speed up the finishing work. Check with your Inspector before attempting to prepare the installation for such a connection.

REQUIREMENTS FOR CONNECTION

To obtain this connection, in most districts the following is required:

Service - Service must be entirely complete, including:

• Meter backing
• Meter blank cover
• Dux seal in last fitting, where required. See p. 25 and p. 27.
• Service grounding
• Waste pipe bonding. Both septic and soap systems, if they are in metal pipe, must be bonded to ground.
• Panel covers must be installed
  Permits - must be complete, that is, they must cover the entire installation, including such appliances as range, water heater, dryer, furnace wiring etc. even though these appliances are not yet installed at the time the temporary connection is required. One branch circuit must be completed by installing all fixtures, plates etc. but leave all other circuits disconnected from their circuit breakers in the panel unless these too are entirely completed. Do not energize any circuit which runs through an

outdoor outlet box unless it is entirely complete with fixtures, fittings, plates, covers etc. It is better to choose a circuit which has all its outlets facing indoors.

In addition, most localities require a temporary connection permit (or TCP) before the Inspector can authorize the power utility to connect the service.

- Building - In some localities it is necessary to close the building for this connection. That is, it must have doors and windows installed before the service can be authorized for connection. Check this with your local Inspector.
- Identify House - The house number should be posted in a conspicuous place on the house. This is to help the Inspector and the power utility connection crew to identify the correct house quickly.
- Completion - When the installation is entirely complete, notify the Inspection department immediately. The Inspector requires written notification (this may be a permit stub properly signed and dated or a special form may need to be filled in.). If the temporary permit expires before the installation is completed, you will need to renew this connection permit.

SPECIAL CASES
Check with your Inspector first.

FOR UNDERGROUND SERVICES ONLY

It is not always necessary to frame the whole building, (i.e. put on the roof and outside sheathing to get a temporary to permanent connection) as described above. Often it is possible to have the building contractor erect only a small portion of the permanent wall where the permanent service (meter base and service panel) is to be located. The permanent meter base and service panel along with one or more plug outlets may then be installed on this partial wall. A temporary wooden enclosure similar to that described for temporary construction service, on "Service Enclosures," p. 130, must be built to protect this electrical equipment from the weather and small prying and inquisitive fingers from getting hurt.

Service Grounding - Permanent service grounding should be installed. This usually consists of 2 - 3 m(118 in.) rods 3 m (118 in.) apart. Don't forget to connect the water piping system when it is installed. See p. 47 for more detailed instructions.

This is only a temporary connection for construction purposes and is subject to immediate disconnection if the terms of use are changed or when the permit expires.

27.   TEMPORARY POLE SERVICES

SECTION 76 IN THE CODE
Confirm these details with your local Inspector.

MINIMUM SERVICE SIZE
Wire size.. #10
Conduit size...3/4 in.
Circuit breaker or Fuse ratings ....15, 20 or 30 amp

POLE REQUIREMENTS
Guide only - check with your Inspector.
- Solid - It must be a pole or solid timber. Laminated timber is not acceptable.
- Size - Minimum thickness is 15.2 cm by 15.2 cm (6 in. by 6 in.) timber, or if a round pole is used, it must be at least Class 6 with a 15.2 cm. (6 in.) diameter top.
- Length - Minimum length is 5.2 m. (17.0 ft.). This provides for 1.2 m. (47.2 in.) in the ground and a minimum 4.0 m. (13. 1 ft.) above ground. This is acceptable only where lines are short and the Hydro pole is on your side of the road. See p. 15 for minimum clearance required for a service line which must cross over buildings, roadways or walkways.
- Gain - Two shallow saw cuts approximately 2 in. apart, with wood chip between removed, marking the pole 3.6 m. (142 in.) from butt end.
- Bracing - as required to offset the pull of the Hydro lines. This should be done with 5 cm x 10 cm. (4 in. x 2 in.) lumber attached as high as possible and should make an angle approximately 30 degrees with the pole.

METER BASE
Height- maximum 185 cm (72.8 in.), minimum 165 cm (64.9 in.), final grade level to center of meter.

Blank Cover - Hydro may require a temporary cover on the base until they can install a meter.

Connection - Connections are as shown on p. 129.

SERVICE EQUIPMENT

Circuit Breaker Type - This is the preferred type of equipment for temporary construction services.

Rule 14-302 requires a double pole breaker with single handle be used. Two breakers with a tie-bar on the handles is not acceptable for this purpose because this is a service switch.

Fuse Type - Rule 14-204 - Non-interchangeable type fuses only may be used. This means that fuse adapters must be installed in the standard fuse sockets in the switch and panel. These adapters are available in different sizes or current ratings. Once a particular adapter has been installed, say a 20 amp size, only a 20 amp fuse will fit this socket.

In some districts only circuit breaker type equipment is permitted. You should check with your local Inspector before proceeding.

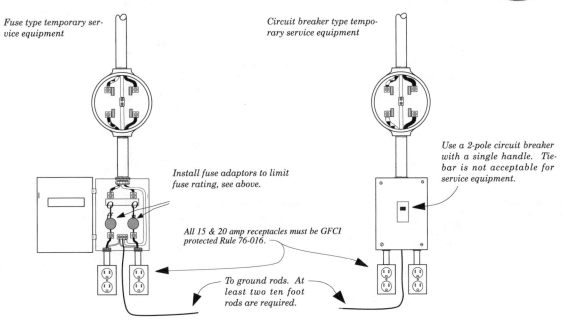

Fuse type temporary service equipment

Circuit breaker type temporary service equipment

Install fuse adaptors to limit fuse rating, see above.

Use a 2-pole circuit breaker with a single handle. Tie-bar is not acceptable for service equipment.

All 15 & 20 amp receptacles must be GFCI protected Rule 76-016.

To ground rods. At least two ten foot rods are required.

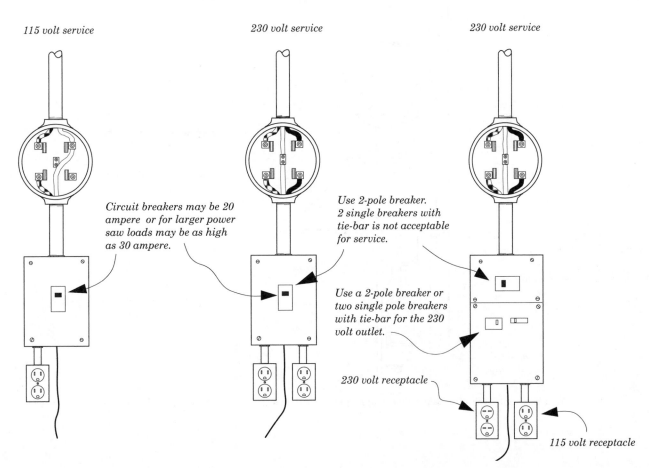

115 volt service

230 volt service

230 volt service

Circuit breakers may be 20 ampere or for larger power saw loads may be as high as 30 ampere.

Use 2-pole breaker. 2 single breakers with tie-bar is not acceptable for service.

Use a 2-pole breaker or two single pole breakers with tie-bar for the 230 volt outlet.

230 volt receptacle

115 volt receptacle

SERVICE ENCLOSURES

Standard Non-Weatherproof Type - Non-weatherproof service equipment may be used if it is installed in a solidly constructed weatherproof box. This box may be of lumber or plywood not less than 2 cm (3/4 in.) in thickness with hinged door and lock. If practicable, the door should be hinged at the top.

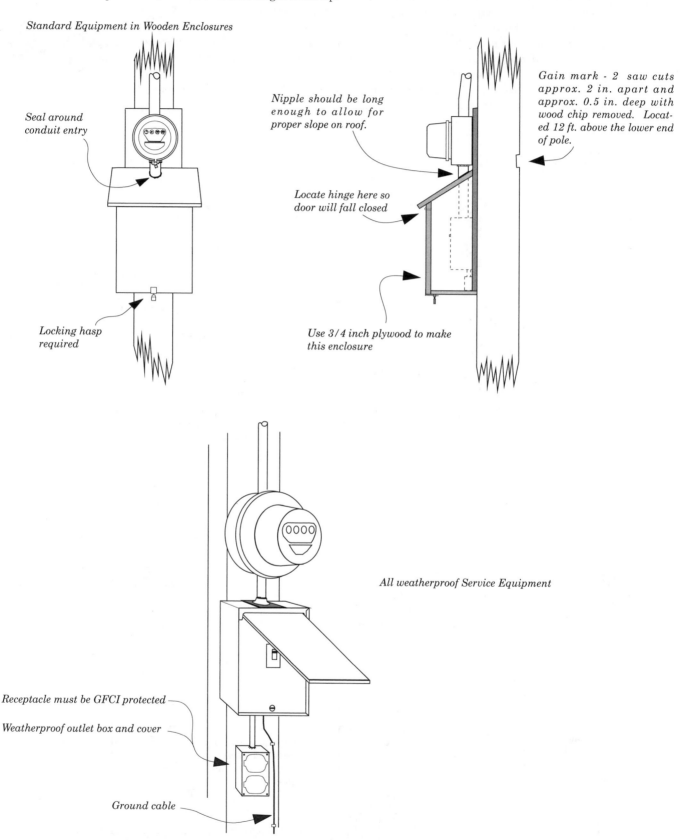

Standard Equipment in Wooden Enclosures

Seal around conduit entry

Nipple should be long enough to allow for proper slope on roof.

Gain mark - 2 saw cuts approx. 2 in. apart and approx. 0.5 in. deep with wood chip removed. Located 12 ft. above the lower end of pole.

Locate hinge here so door will fall closed

Locking hasp required

Use 3/4 inch plywood to make this enclosure

All weatherproof Service Equipment

Receptacle must be GFCI protected

Weatherproof outlet box and cover

Ground cable

Weatherproof Type - It is better than standard equipment in a wooden enclosure. It is also much simpler to install.

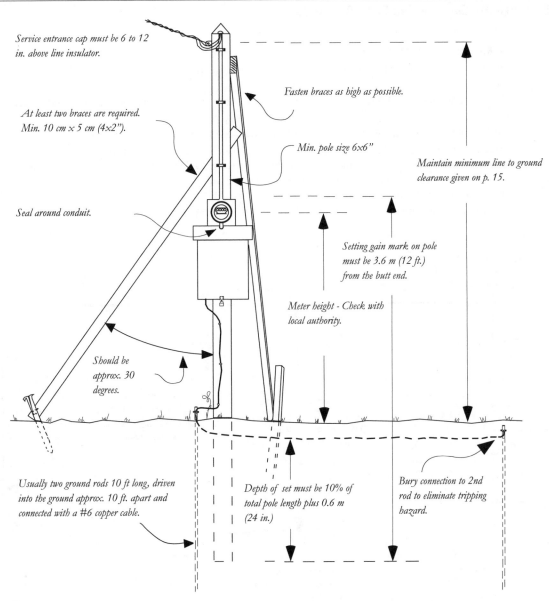

Service entrance cap must be 6 to 12 in. above line insulator.

Fasten braces as high as possible.

At least two braces are required. Min. 10 cm x 5 cm (4x2").

Min. pole size 6x6"

Maintain minimum line to ground clearance given on p. 15.

Seal around conduit.

Setting gain mark on pole must be 3.6 m (12 ft.) from the butt end.

Meter height - Check with local authority.

Should be approx. 30 degrees.

Usually two ground rods 10 ft long, driven into the ground approx. 10 ft. apart and connected with a #6 copper cable.

Depth of set must be 10% of total pole length plus 0.6 m (24 in.)

Bury connection to 2nd rod to eliminate tripping hazard.

RECEPTACLES

120 Volt Receptacles - A 2 wire, 120 volt service would have a single pole circuit breaker. It could serve one 120 volt plug receptacle.

A 3-Wire, 120/240 Volt Service would have a 2-pole circuit breaker. It could serve two 120 volt plug receptacles or one 240 volt receptacle without requiring a branch circuit panel.

240 Volt Receptacles - Whenever a 240 volt plug outlet is required, in addition to a 120 volt outlet, you must install a branch circuit panel. A combination circuit breaker panel is the best way out of this awkward requirement.

GROUNDING

Cable - must be of copper. Where the service is 100 amp or less the grounding conductor may be #4 bare or even #6 bare copper if not subject to mechanical damage. See p. 45 for details.

Ground Rods - Use only manufactured grounding electrodes. Both rods and plates must be certified by CSA or by one of the other approved certification agencies, see p. 2.

Copper rods... 12.7 mm (1/2 in.) by 3 m (118 in.)
Solid iron rod ....15.8 mm (5/8 in.) by 3 m (118 in.)

Use solid rods - Driven pipe is not approved for grounding, Rule 10-700.

How Many Rods? - Rule 10-700(2)(a) - In most localities 2 rods driven 3 m (approx. 10 ft.) apart are required. In others you may need more to provide adequate grounding for the service.

28.   PRIVATE GARAGE AND FARM BUILDINGS

OVERHEAD SUPPLY LINES

Insulation of Wire - Rule 12-302 - Conductors must have weather proof insulation. Triplex cable is normally used today. Rule 12-318.

Elevation of Wire - Rule 12-304 - Overhead lines must be out of reach and have the line to ground clearances given on p. 15.

Roof Crossing - Overhead lines may cross a roof provided the minimum line clearances given on p. 15 are maintained.

Triplex Type Cable - Rule 12-318 - This consists of two insulated conductors wrapped around a bare messenger cable. This cable can be used to bring two 15 amp circuits to the outbuilding. In that case the bare messenger would serve as a neutral for the two circuits and grounding would be as shown. Such an installation could also be used for a building housing livestock. In some cases the bare messenger, in triplex cable, may be used as the bonding conductor between the two buildings.

Two 15 amp. circuits may be run overhead as shown but if more circuits are needed a sub-panel is required in the out-building.

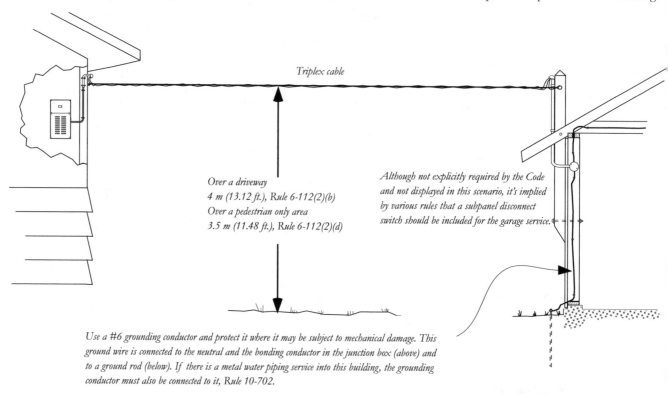

Triplex cable

Over a driveway
4 m (13.12 ft.), Rule 6-112(2)(b)
Over a pedestrian only area
3.5 m (11.48 ft.), Rule 6-112(2)(d)

Although not explicitly required by the Code and not displayed in this scenario, it's implied by various rules that a subpanel disconnect switch should be included for the garage service.

Use a #6 grounding conductor and protect it where it may be subject to mechanical damage. This ground wire is connected to the neutral and the bonding conductor in the junction box (above) and to a ground rod (below). If there is a metal water piping service into this building, the grounding conductor must also be connected to it, Rule 10-702.

The insulated wire holder at each end must be single spool type such as shown. Do not use lag screw. This thing must be held in place with a bolt through the wall or through the mast as shown.

Grounding - Rule 10-204 - The grounded circuit conductor must be bonded to ground at each building. A #6 grounding conductor may be used as shown, where it is not subject to mechanical damage, Rule 10-806(6). The connection may be made in the first junction box. Do not use solder for this connection. Normally, only one ground rod is required for each such sub-service.

UNDERGROUND SUPPLY LINES

RULE 12-012

Cable Type —is NMWU - Table 19 (This is the old NMW10)

Cable Size — is #14 but for long runs, above 15 m. (50 ft.), #12 is recommended.

Cable Depth - without a protecting plank above the cable
• 60 cm (23.6 in.) under pedestrian only area.
• 90 cm (35.4 in.) under vehicular traffic areas.

Cable Depth - With a protecting plank above the cable as shown below.
• 45 cm (17.7 in.) under pedestrian only area.
• 75 cm (29.5 in.) under vehicular traffic areas.

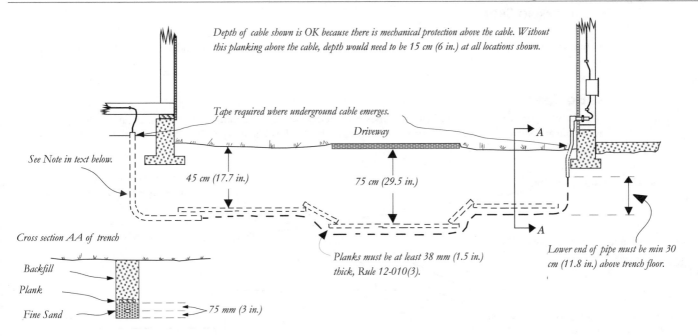

Depth of cable shown is OK because there is mechanical protection above the cable. Without this planking above the cable, depth would need to be 15 cm (6 in.) at all locations shown.

Tape required where underground cable emerges.

Driveway

A

See Note in text below.

45 cm (17.7 in.)

75 cm (29.5 in.)

A

Cross section AA of trench

Backfill

Plank

Fine Sand

75 mm (3 in.)

Planks must be at least 38 mm (1.5 in.) thick, Rule 12-010(3).

Lower end of pipe must be min 30 cm (11.8 in.) above trench floor.

Cable Protection - Conduit should extend into the trench as shown and cables underground should be protected as follows:

- 75 mm (3 in.) of sand below cable
- 75 mm (3 in.) of sand above cable.
- 38 mm (1.5 in.) plank above the sand —not nominal, but actual thickness.

Polyethylene water pipe is usually permitted provided we use it for the underground portion only. The riser at each end must be either rigid PVC or rigid metal conduit, not EMT.

SUB-PANELS (GARAGE OR SIMILAR BUILDING)

Where the load in the garage or other separate building requires more than two 15 amp circuits, or where you wish to provide for future load additions, a sub-panel may be installed.

SIZES

Where plug outlets are used to supply power tools in a private garage or workshop it is recognized that there will normally be only one person working these tools at any one time. Therefore, even if you have a number of large power tools it is unlikely that more than one such tool would be used at any one time. For this reason, when we are calculating the feeder size for this panel, we need to concern ourselves with only one machine, the one with the largest electrical load then add all the other small loads.

Wood and Metal Lathe

#10 SUB-FEEDER

If there is no fixed electrical heating load and no electric welder load a #10 NMD90 (copper) cable may be sufficient. This cable could supply a 30 ampere load at 240 volts. It could supply a 1-1/2 HP power tool at 240 volts, one circuit for lighting and one for plug outlets. The panel should be large enough for present loads plus some reserve capacity for future load additions.

Drill Press

Power Saw

LARGER LOADS

Such as welders or fixed electric heaters would require more power and therefore a larger cable. A #8 NMWU (copper) cable will supply 40 amperes at 240 volts. This is large enough to supply a small private garage type welder, one circuit for lighting and one or more circuits for plug outlets. If the full load current of the welder was say 20 amps, then the circuit breaker supplying the 40 ampere NMWU sub-feeder cable could be as high as 60 amperes. Obviously, the load may not be greater than 40 amperes on this 40 ampere cable but because the biggest load is either a welder load or a motor load the code permits the ampere rating of the main supply breaker to be greater than the cable rating.

WOODWORKING SHOP

A 40 ampere NMWU copper feeder cable supplied with a 60 ampere circuit breaker in the main panel could serve a 3 HP 240 volt motor in the shop. There would be sufficient capacity left to supply garage lighting and a number of branch circuits for plug outlets.

THE BIG ONE

A #6 NMWU (copper) feeder to a garage or workshop is better but it may be more than is necessary in most cases. If you have an unusually large load, or the length of this feeder to the 2nd building is say 80 ft or more, you should consult your local Electrical Inspector for advice.

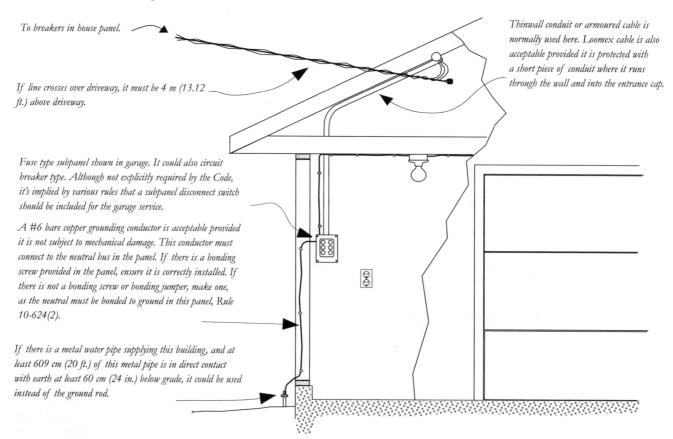

To breakers in house panel.

If line crosses over driveway, it must be 4 m (13.12 ft.) above driveway.

Thinwall conduit or armoured cable is normally used here. Loomex cable is also acceptable provided it is protected with a short piece of conduit where it runs through the wall and into the entrance cap.

Fuse type subpanel shown in garage. It could also circuit breaker type. Although not explicitly required by the Code, it's implied by various rules that a subpanel disconnect switch should be included for the garage service.

A #6 bare copper grounding conductor is acceptable provided it is not subject to mechanical damage. This conductor must connect to the neutral bus in the panel. If there is a bonding screw provided in the panel, ensure it is correctly installed. If there is not a bonding screw or bonding jumper, make one, as the neutral must be bonded to ground in this panel, Rule 10-624(2).

If there is a metal water pipe supplying this building, and at least 609 cm (20 ft.) of this metal pipe is in direct contact with earth at least 60 cm (24 in.) below grade, it could be used instead of the ground rod.

GROUNDING

Rule 10-208(a) - Says the grounded circuit conductor must be bonded at each building. This has nothing whatever to do with the question of livestock in the building. This is service neutral grounding and it is required at every building without exception. Subrule (b) does not apply to the grounded neutral, Note that it refers only, and specifically, to the non current carrying metal parts, such as the metal conduit, the metal outlet boxes etc. but does not refer to the neutral conductor. This Subrule allows these non current carrying metal parts to be bonded back to the main service panel although it is difficult to see why anyone would want to do this. Subrule (b) will permit a separate bonding conductor to be run with the supply conductors back to the main panel if the building is not used for livestock. This is strictly bonding of electrical equipment and if we provided this bonding conductor we would still be required to ground the neutral as shown. The point here is that in every case the neutral must be grounded at every building. See also Rule 10-204.

A #6 grounding conductor may be used if it is not subject to mechanical damage, Rule 10-806. This grounding conductor must be connected to the neutral bus in the branch circuit panel as shown. If the neutral in that branch circuit panel is not already bonded to the enclosure you will need to bond it as shown. This is a very important safety concern.

Ground Rods - Normally, only one 3 m (10 ft.) ground rod is required for each sub-service panel in a separate building.

EMT feeder, see illustration above.

Connect the black & red wires to the buses. The white wire connects to the neutral bus.

The grounding conductor loops through this connector lug then continues on to the neutral bus

29.   DUPLEX DWELLINGS

SEPARATE SERVICE CONDUCTORS

Rules 6-104 & 6-200 - In duplex dwellings each suite may be served with separate service equipment as shown. The meter base, conduit sealing & draining, the service panel and all the instructions for grounding would be exactly the same as that given in the appropriate sections in this book for a single family residence.

The two entrance caps must be located close together so that only one Hydro service drop is required.

COMMON SERVICE CONDUCTORS

Rule 6-200(2) - This rule permits a two-gang meter base to be used in a duplex dwelling without a main switch ahead of it.

The main service conductors must be large enough to carry 100% of the calculated load in the heaviest loaded suite plus 65% of the calculated load in the other suite plus the heating and air conditioning loads. See sample calculation below.

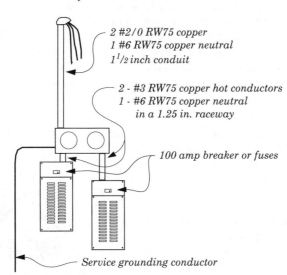

2 #2/0 RW75 copper
1 #6 RW75 copper neutral
1^1/2 inch conduit

2 - #3 RW75 copper hot conductors
1 - #6 RW75 copper neutral
    in a 1.25 in. raceway

100 amp breaker or fuses

Service grounding conductor

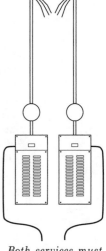

Both services must be grounded to the same grounding electrode.

CALCULATION

If both units of the duplex are to be served from a common service the size of the conductors to each suite should be as follows:

EXAMPLE OF A CALCULATION - IDENTICAL LOAD IN EACH SUITE

Basic Load 102.2 m^2 (1,100 sq. ft.) floor area

First 90 m^2 (968.4 sq. ft.)	5,000 watts
Next 12.2 m^2 (131.6 sq. ft.)	1,000 watts
10 kw. range	6,000 watts
4 kw. dryer	1,000 watts
3 kw. water heater	750 watts
Total	13,750 watts
Electric heating	7,000 watts
	20,750 watts

$$\frac{20,750}{240} = 86.4 \text{ amps}$$

Each suite requires 100 amp service panel with 100 amp breaker.

2 - #3 RW75 copper wire in 1-1/4 in. suite service conduit.
1 #6 RW75 copper conductor - white, for the neutral
Use 1 in. (27) or 1-1/4 in. (35) for suite service conduit.

MAIN SERVICE SIZE CALCULATION

13,750 watts at 100% ....13,750 watts

13,750 watts at 65% ....8,937 watts

Heating total is 14,000 watts

10,000 at 100% .... 10,000 watts

4,000 at 75% ....3,000 watts

_____

35,687 watts

$$\frac{35,687}{240} = 148 \text{ amps total calculated load}$$

Main service size is 148 amps

Metering and Service Conductors - Normally the suite service panels can be installed within a few ft. of the meter base, as shown above. Rules 6-206(1)(e) & 6-208 limit service conductor length inside the building to "as close as practicable." This means just through the wall, about 10 in. long. See p. 20 for some exceptions to this minimum length inside a building. On the outside of the building these runs may be a bit longer but the total length of each individual suite feeder must not exceed 24.6 ft (7.5 m). This is measured from the meter base terminal to the suite service breaker, it is the actual length of that feeder. The reason for this limit of 24.6 ft. is that these suite feeders are smaller than the main service conductors, see Rule 14-100(c).

Service Equipment - The service equipment in each suite normally consists of a combination service panel as shown.

Service Grounding - In this case the service grounding conductor must run into the duplex meter base. These meter bases usually have provision for the grounding conductor connection and for bonding the neural to the meter enclosure. See "Grounding and Bonding," p. 45.

Size of Grounding Conductor - may be determined from the List of Materials, p. 5.

The size of this grounding conductor is related to the current carrying capacity of the main service conductor, use Table #17.

30.   REWIRING AN EXISTING HOUSE

ELECTRICAL SERVICE PANEL

If you are adding load to an existing panel in an old house but have no more breaker or fuse spaces you will need to install an additional sub panel and feed it from breakers in the existing panel. Before you do this, make sure the service equipment is adequate to supply the new load. Use the table on p. 18 to determine size of service needed for all the load. If there is sufficient spare capacity, reroute two of the existing circuits to a new panel located nearby. This will free up space for two breakers for a sub-feeder to supply the new panel.

CHANGE SERVICE PANEL

This will cost a bit more but it may be the only solution if the old stuff is too small. Don't forget that if you change the panel you must upgrade the whole service, not the branch circuit wiring but the service. The reason for this is the new service may require larger conductors, conduit, meter base and grounding. It may also be necessary to find a more acceptable location for the new panel, meter base or service entrance cap. It may also be that all of these items are satisfactory and all that is needed is to replace the old panel with a larger one.

All this may sound expensive and indeed may be but don't forget this is the most important part of your installation. This is what protects all the circuits in the house and this is where that enormous hydro electric capacity is limited to a safe value. To compromise here is to compromise with minimum safety.

Even if it looks harmless, Don't take chances. Before any work is begun you should make sure that part of the system you will be working on is, in fact, not energized. If you are changing your service equipment have Hydro crew disconnect your service before you begin work.

Where a service has been changed the Inspector will usually check for:

- Entrance cap - Re: height and location, see "Hydro Service Wires to House," p. 14.
- Size of Service Conduit & Conductors - This may have been large enough at one time but may be too small now. See p. 4.
- Meter Location etc. - Accessibility and distance from the front of the house, see p. 14.
- New Panel - Must comply with new Code re: ampacity, number of circuit positions and number of 2-pole positions available for 240 volt and 3 wire circuits, p. 5 and p. 41.
- Service Grounding - Size of conductor, condition of run, connected to old abandoned water service? Must comply with current regulations, see p. 45.
- Service Bonding - Rule 10-406(3) - Older houses may have metal waste pipe system which may not have been bonded. In that case the Code requires a bonding connection between the metal waste piping and the service grounding electrode, or to the nearest bonded cold water pipe.

BRANCH CIRCUIT WIRING

ADDITIONS OR CHANGES TO THE HOUSE

All new wiring must comply with current regulations. The information on branch circuit wiring, given in this book, does apply to any new additional wiring.

Existing wiring - The Inspector will not normally require the existing finished areas to be rewired except as may be necessary for minimum safety. They will, however, require upgrading of branch circuit wiring where the walls have been opened to make other structural revisions or additions to the house. They have the knowledge and the experience to guide them in their assessment of your installation. They will ask for minimum changes to the existing wiring to make your home reasonably safe. The fact is - it's your safety they are concerned about and you want them to be concerned about that.

Where your new cable is used in concealed wiring and it impractical to provide supports, if it is running through metal joists, metal top or bottom plates, or metal sheeting or cladding, it`s permitted to fish the cable. I know, it`s unlikely that the inner-city nineteen-teen CPR special you`re retrofitting will have metal studs concealed behind the horse-hair lathe and plaster, but the Code is making concessions nonetheless.

Replacing Old Plug Receptacles - Subrules 26-700(7)(8) & (9) - If you are fixing up an old house you may find the branch circuit wiring is with single insulated conductors supported on insulator knobs, (this is called knob and tube wiring) or, it may be the wiring consists of old type loomex cable which did not have a bonding conductor in the cable. In either case all the plug outlets will be ungrounded type. The Code permits us to use grounded type receptacle in these ungrounded outlet boxes provided we do one of the following things:

• Install the new grounded type plug receptacle, (it could be a GFCI type but does not need to be) and bond it as shown in the illustration. The illustration shows an outlet in a finished wall where you may need to fish a new bonding conductor down through the floor into the basement area where connection can be made to a grounded cold water metal pipe. A #14 solid copper conductor with green insulation may be used. It must be installed where it will not be subject to mechanical damage. The clamp must be an approved ground clamp. A copper ground strap arrangement is not sufficient; or

Use 2 bolts - do not try to do this with one bolt.

This strap may be used for bonding gas tubing to the service ground.

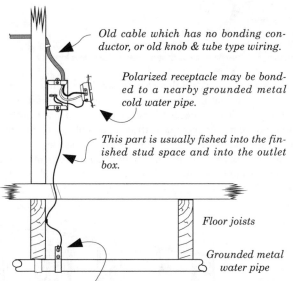

Old cable which has no bonding conductor, or old knob & tube type wiring.

Polarized receptacle may be bonded to a nearby grounded metal cold water pipe.

This part is usually fished into the finished stud space and into the outlet box.

Floor joists

Grounded metal water pipe

This bonding conductor is usually fished into the basement where it is connected to a nearby, grounded cold water pipe.

• Install a GFCI type circuit breaker in the panel then use grounded type receptacles in all the outlets on this circuit as shown in the top circuit in the illustration below; or

• Install a GFCI type receptacle as shown in the lower circuit below. Make sure the connections are correct so that all the other downstream receptacles will also have GFCI protection. The illustration shows new additional outlets added to this existing circuit. In some difficult cases this may be a good solution but the rule does not clearly say it may be so done, therefore, you should check first with your Inspector.

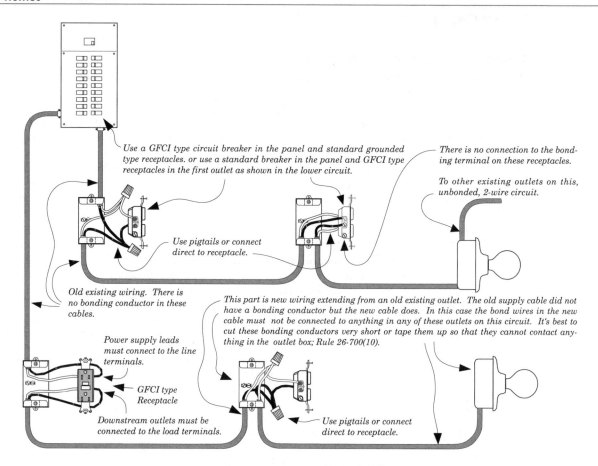

Use a GFCI type circuit breaker in the panel and standard grounded type receptacles. or use a standard breaker in the panel and GFCI type receptacles in the first outlet as shown in the lower circuit.

There is no connection to the bonding terminal on these receptacles.

To other existing outlets on this, unbonded, 2-wire circuit.

Use pigtails or connect direct to receptacle.

Old existing wiring. There is no bonding conductor in these cables.

This part is new wiring extending from an old existing outlet. The old supply cable did not have a bonding conductor but the new cable does. In this case the bond wires in the new cable must not be connected to anything in any of these outlets on this circuit. It's best to cut these bonding conductors very short or tape them up so that they cannot contact anything in the outlet box; Rule 26-700(10).

Power supply leads must connect to the line terminals.

GFCI type Receptacle

Downstream outlets must be connected to the load terminals.

Use pigtails or connect direct to receptacle.

Part of the lower circuit in the illustration above is new work, i.e. these are new outlets extended from an old existing circuit. Subrules 26-700(7)(8) & (9) do not refer to new extensions to old existing installations, they refer only to existing outlets. However, Subrule (9) says there must not be any bonding connection between plug receptacles unless the bonding connection is complete all the way back to the panel. In this case it is not complete, the supply cable to the first outlet box has no bonding conductor, therefore, the bond wires in all of the outlet boxes must be left completely disconnected from the boxes, the receptacles and from other bond wires. Crazy? No, not really. When the bonding conductor connects these receptacles together a fault on one outlet will also appear on all the other outlets connected together and that could be very dangerous.

Electrical Permit - make sure your electrical permit is adequate to cover the work done.

31.   MOBILE HOMES

CSA MARKING

The mobile unit must have a CSA label or a label applied by one of the other certification agencies noted on p. 2. If your unit does not have such a label you are in deep trouble. Sometimes it is lost in the move or for one reason or another it is gone. This approval label is one of the first things the Inspector will look for. If it is not there they cannot

authorize a service connection. Your local inspection office can advise you on the procedure for replacing a lost certification label.

ELECTRICAL PERMIT

An electrical permit must be obtained before any electrical work is done.

If this mobile unit is on private property and you are personally going to occupy it, you are usually permitted to obtain a permit and do the work; or

If this mobile unit is to be occupied by someone in your immediate family, you are usually permitted to obtain an electrical permit and do the work, but if the unit is for rental purposes to someone other than immediate family, the wiring must be installed by an electrical contractor. And, if this unit is to be connected in a mobile home park, the work must be done by a certified (licensed) electrical contractor. The owner would not normally be permitted to do this work nor to take out a permit.

SERVICE SIZE

Rule 72-102(1) - Mobile homes are treated the same as a single family dwelling as far as service ampacity is concerned. See the table on p. 18 to determine the required ampacity of service equipment.

Rule 8-200, Floor area - less than 80 m^2 (861 sq. ft.)

The normal load will be something like this.

Basic..5,000 watts
Range - (12 kw. range) ....6,000 watts
Water heater - (3 kw tank)..750 watts
Dryer — (4 kw. unit)..1,000 watts
Quick Steam Tap — (500 watts) ....0 watts
Furnace — (gas or oil)...0 watts
Freezer...0 watts
Dishwasher..0 watts
Fridge..0 watts

Total	12,750 watts
12,750 divided by 240 volts	= 53.1 amps.

The code requires a 60 amp service for this load based on floor area. Also check Table 39 on p. 5 to see whether a lower ampacity is sufficient.

It really does not matter if the range, dryer, water heater and furnace are all gas - the fact is, the minimum service size is still 60 amp for any size floor area up to 80 m^2.

Larger Floor Area - Where the floor area is 80 m^2 (861 sq. ft.) or larger the minimum size service acceptable by Code is 100 amps.

Unfortunately, the rules make no distinction between a single family house and a mobile home. The same rules used to determine the size of a service in a house are also applied in determining the size of service in a mobile home.

We all know that the calculated load in this unit is much less than 100 amp The extra service ampacity is usually intended for future load changes. In single family houses there is always the possibility of finishing a basement, adding a sauna or large power vacuum etc. In a mobile home such changes are difficult, very costly and rare. But the minimum service size is the minimum service size - it's the law.

SINGLE UNIT IN MOBILE PARK

CONNECTION BOX IN MOBILE HOME PARK

A typical connection box in each lot in a mobile home park.

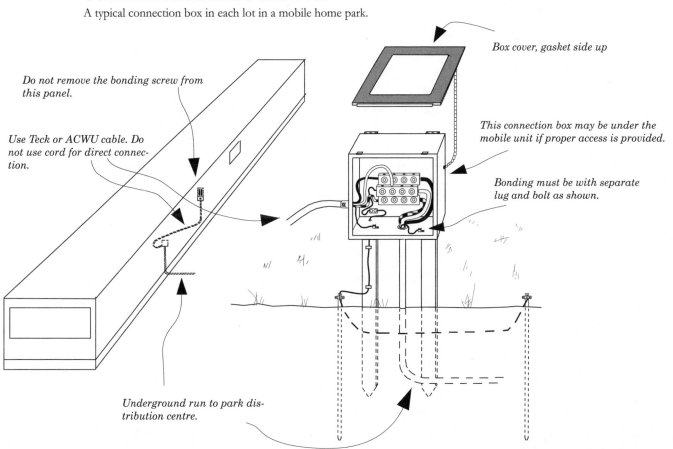

Box cover, gasket side up

Do not remove the bonding screw from this panel.

Use Teck or ACWU cable. Do not use cord for direct connection.

This connection box may be under the mobile unit if proper access is provided.

Bonding must be with separate lug and bolt as shown.

Underground run to park distribution centre.

Connection to this box can only be made by a certified electrical contractor. Neither the owner of the mobile unit nor the mobile park operator (or owner) may make this connection unless they have a valid Certificate of Qualification to do this work. An electrical permit must be obtained before connection is made.

Cable Connector - The cable from the mobile unit must connect to the box with a connector approved for the particular cable you are using.

Conductor Terminations in Box - Watch this carefully - most of the problems are right here.

- Don't skin back too much insulation - only enough to make good connections.
- If the box does not have a terminal strip as shown - use only approved connectors to make the splices. This is an extremely important detail when aluminum conductors are used. In that case use only connector lugs marked approved for aluminum and make sure the conductor surface has been cleaned and an inhibitor has been applied.
- Bonding - Terminate the grounding conductor in an approved lug which is separately bolted to the box. Don't try to use a bolt or screw already used for some other purpose.
- Clean the Surface under the bonding lug to ensure a good electrical contact.

Single Unit on Private Property

Basic Requirements

Permission to Place Mobile Home - More and more of our freedoms are being taken away from us. Before you move a mobile unit onto your own property, it is best to find out first if that would be permitted by local by-law. In some cases you will be required to produce some kind of proof of approval from the local authorities before the Electrical Inspection department will issue an electrical permit for connection.

Consult Hydro - Rules 6-206(1) & 6-114 - Before any work is done the power utility should be consulted to determine which pole service will be from.

Connection Methods

There are several connection methods you may choose from - each has its problems, pitfalls, and advantages.

#1 Dip Service

As shown below, only the meter is on the pole. The service conductors dip underground to a point under the panel where they rise and enter the service box in the mobile unit. This means the service conductors do not have fuse protection until they enter the panel. It means there is no quick easy shut-off, except by removal of the meter when the moving truck arrives.

Meter only on service pole.

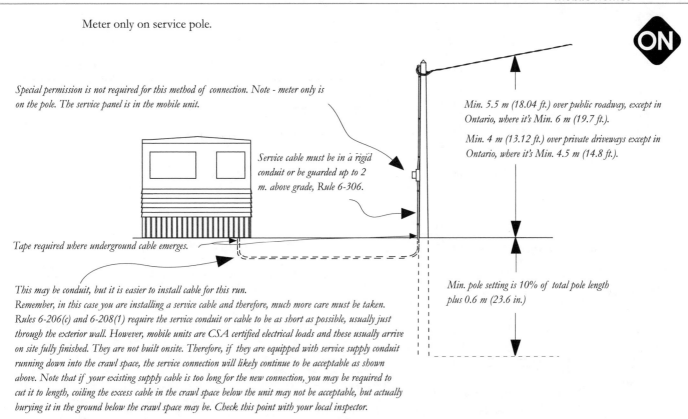

Special permission is not required for this method of connection. Note - meter only is on the pole. The service panel is in the mobile unit.

Service cable must be in a rigid conduit or be guarded up to 2 m. above grade, Rule 6-306.

Tape required where underground cable emerges.

This may be conduit, but it is easier to install cable for this run. Remember, in this case you are installing a service cable and therefore, much more care must be taken. Rules 6-206(c) and 6-208(1) require the service conduit or cable to be as short as possible, usually just through the exterior wall. However, mobile units are CSA certified electrical loads and these usually arrive on site fully finished. They are not built onsite. Therefore, if they are equipped with service supply conduit running down into the crawl space, the service connection will likely continue to be acceptable as shown above. Note that if your existing supply cable is too long for the new connection, you may be required to cut it to length, coiling the excess cable in the crawl space below the unit may not be acceptable, but actually burying it in the ground below the crawl space may be. Check this point with your local inspector.

Min. 5.5 m (18.04 ft.) over public roadway, except in Ontario, where it's Min. 6 m (19.7 ft.).

Min. 4 m (13.12 ft.) over private driveways except in Ontario, where it's Min. 4.5 m (14.8 ft.).

Min. pole setting is 10% of total pole length plus 0.6 m (23.6 in.)

#2 MAST ON MOBILE UNIT

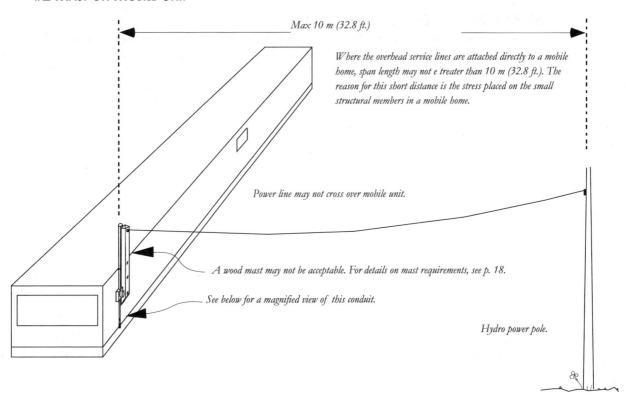

Max 10 m (32.8 ft.)

Where the overhead service lines are attached directly to a mobile home, span length may not e treater than 10 m (32.8 ft.). The reason for this short distance is the stress placed on the small structural members in a mobile home.

Power line may not cross over mobile unit.

A wood mast may not be acceptable. For details on mast requirements, see p. 18.

See below for a magnified view of this conduit.

Hydro power pole.

It is often difficult to provide adequate mast support on a mobile unit. For this reason length of this overhead service drop should be as short as possible.

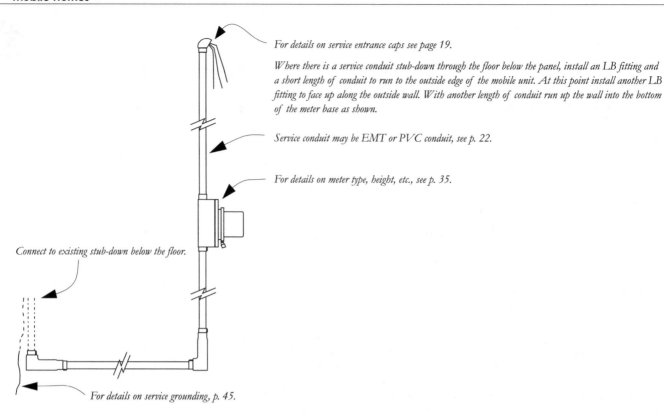

For details on service entrance caps see page 19.

Where there is a service conduit stub-down through the floor below the panel, install an LB fitting and a short length of conduit to run to the outside edge of the mobile unit. At this point install another LB fitting to face up along the outside wall. With another length of conduit run up the wall into the bottom of the meter base as shown.

Service conduit may be EMT or PVC conduit, see p. 22.

For details on meter type, height, etc., see p. 35.

Connect to existing stub-down below the floor.

For details on service grounding, p. 45.

The manufacturer of the mobile unit normally provides a service conduit stub down through the floor directly below the panel. Install an L.B. on this stub down and a short length of conduit to run to the outside edge of the mobile unit as shown in the illustration. At this point install another L.B. to face up on the outside wall and with another length of conduit run up the wall into the bottom of the meter base.

#3 POLE SERVICE - BY SPECIAL PERMISSION ONLY

This is an outdoor service and Rule 6-206(1)(c) says special permission is required for such an installation. Special permission for these installations must be obtained before any work is done. Service ampacity is based on Rule 72-102(1) which is the same as for a house of similar size. See the table on p. 18.

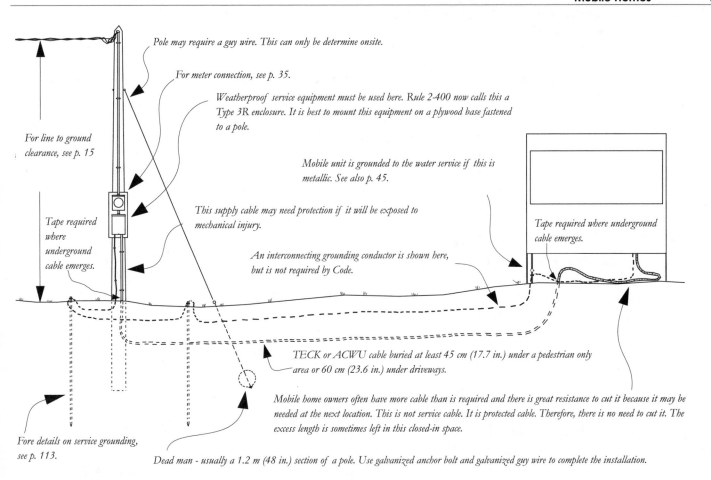

Pole may require a guy wire. This can only be determine onsite.

For meter connection, see p. 35.

Weatherproof service equipment must be used here. Rule 2-400 now calls this a Type 3R enclosure. It is best to mount this equipment on a plywood base fastened to a pole.

For line to ground clearance, see p. 15

Mobile unit is grounded to the water service if this is metallic. See also p. 45.

Tape required where underground cable emerges.

This supply cable may need protection if it will be exposed to mechanical injury.

Tape required where underground cable emerges.

An interconnecting grounding conductor is shown here, but is not required by Code.

TECK or ACWU cable buried at least 45 cm (17.7 in.) under a pedestrian only area or 60 cm (23.6 in.) under driveways.

Mobile home owners often have more cable than is required and there is great resistance to cut it because it may be needed at the next location. This is not service cable. It is protected cable. Therefore, there is no need to cut it. The excess length is sometimes left in this closed-in space.

Fore details on service grounding, see p. 113.

Dead man - usually a 1.2 m (48 in.) section of a pole. Use galvanized anchor bolt and galvanized guy wire to complete the installation.

Grounding - Use two 3 m. (10 ft.) ground rods and connect to the metal water service pipe under the unit. See p. 45 for details.

Cable Types - ACWU, TECK and Corflex may be used. NMWU may also be used when provided with adequate mechanical protection - not recommended for this application.

Service - See index for detailed list of requirements.

POLE LOCATION ON PROPERTY

This is very important — consider the following:

(1). Near Connection Point - The pole should be (not must be) located near the point of service connection on the mobile unit. The length of your existing supply cable will often determine this.

(2). Length of Hydro Lines Required - Note the Hydro power pole location. Hydro will usually run 30 m (98 ft.) onto your property without additional fee. Beyond this length they may be into your wallet for more money. Long runs, more than 30 m. (98 ft.) may also require additional poles and this is at your expense, usually.

(3). Crossings
- Public Roadway- Rule 6-112(2)(b) - You may not have a choice. If the Hydro line is on the other side of the road you must cross the road with the line. Clearance required is 5.5 m. (18.04 ft.) minimum above roadway. It may require a long pole to do this.
- Driveway - Avoid crossing driveways if possible. If you must cross, the minimum line to ground clearance is 4 m. (13.12 ft.) above a driveway. This is generally considered the minimum height even if there is no garage or carport, only a driveway.
- Roof - Do not cross the roof of any building including the roof of the mobile unit unless you can provide the clearances required by Rule 12-310. See p. 15 for a simplified list of required clearances.

POLE REQUIREMENTS

Check with your Inspector.

Ontario amendments allow that secondary and primary line poles be made of wood, steel, concrete, fibre-reinforced polymer, or other acceptable material.

Types - Cedar, fir, lodge pole pine are usually acceptable. If you go into the woods to cut down a tree to make a pole, observe the following picky points:
- Make sure it is straight and has the correct dimensions - see below under 'length'.

- Remove all branches, leaving a smooth surface.
- Remove all bark.
- Cut a gain mark 3.6 m. (14.2 in.) from butt, as shown below.
- Apply an effective treatment to both ends to retard rotting.

It may be possible to use a sawn timber instead of a round pole. It is not recommended and before you try it, check with your Inspector.

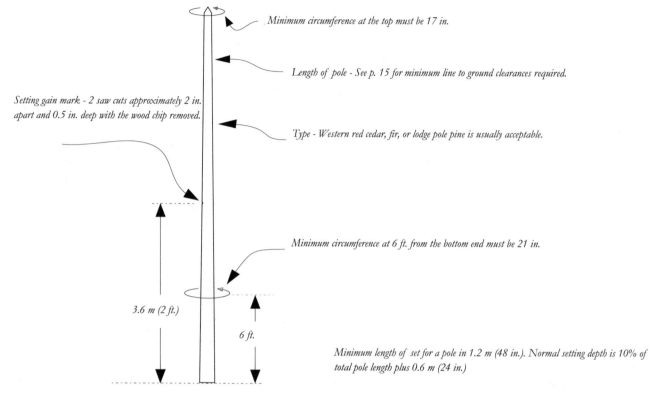

Minimum circumference at the top must be 17 in.

Length of pole - See p. 15 for minimum line to ground clearances required.

Setting gain mark - 2 saw cuts approximately 2 in. apart and 0.5 in. deep with the wood chip removed.

Type - Western red cedar, fir, or lodge pole pine is usually acceptable.

Minimum circumference at 6 ft. from the bottom end must be 21 in.

3.6 m (2 ft.)

6 ft.

Minimum length of set for a pole in 1.2 m (48 in.). Normal setting depth is 10% of total pole length plus 0.6 m (24 in.)

Size - Must be minimum Class 6 pole. This means it must be:
- Minimum 43 cm (17 in.) circumference at the top of the pole. This is minimum, it may be larger.
- Minimum 70 cm (27.6 in.) circumference 1.8 m (6 ft.) from the butt end for a 7.7 m (25 ft.) pole. Longer poles must be even larger in circumference.

Length - Two factors must be considered.
- Length of pole above ground. This is simply line to ground clearances given on p. 15. Add to this the expected sag in your line, AND then add:
- Length of pole set into the ground. In most cases this is 1.2 m (47.2 in.). Where the pole length above ground must be greater than 6 m (19.7 ft.) the length of pole set into the ground must be increased. The normal setting in depth is 10% of total pole length plus 60.96 cm (24 in.).

Guy Lines - When Required? This must be determined on site. A very rough thumb rule is as follows:
- If the length (of pole above ground) is 4.5 m (14.75 ft.) or less and the length of span for the Hydro lines is say 10 m. (59 ft.) a guy is likely not required. If the dimensions are greater than these, it is very likely you will need to guy the pole.
- Type of Guy Wire - must be galvanized cable. Do not try to use ordinary wire rope or clothes line cable.
- Size - 5/16 in. guy wire would be sufficient in most cases.

- Insulator - Where a guy line insulator is required it must be located high enough above ground so that it is out of reach in any position. In Ontario, not withstanding Rule 75-310, guys directly attached to steel poles don't require a guy insulator.
- Dead Man - This is usually a 4 ft. length of log buried to the same depth as the pole setting. Use a long galvanized anchor bolt as shown below.